BRANIFF

INTERNATIONAL AIRWAYS

The Texan Giant 1948–1982

Theodore Gianna

CLASSIC AIRLINES PUBLICATIONS

Front cover photo courtesy of George Mihalek.
Back cover photograph: Braniff International 727. 1968. By Clipperarctic.
https://commons.wikimedia.org/wiki/File:Braniff_International_727_(6961711298).jpg

ISBN: 978-0-6489379-1-3 (Paperback)

First printing edition 2020.

DEDICATION

I DEDICATE THIS BOOK TO ETHAN

CONTENTS

ACKNOWLEDGMENTS

For an exceptionally long time, I had wanted to write a book on my favorite subject, aviation. To be more specific, I wanted to write a history of an airline. I chose Braniff. Why? Because it was one of the airlines that revolutionized air travel. It was the first airline to introduce a new, and flamboyant style which epitomized the 1960s. Braniff's style was one that made America wake up and take notice. They set the pace in forming an image which became an indelible part of American civil aviation history. It was an airline that was ahead in the fashion stakes in the glorious 1960s, with its artistic use of colors. Its "hostesses" which were specially chosen for their looks and vitality, and all of Braniff's staff were exuberantly helpful, and attentive. As for its offices, and airport waiting areas, they were decked out with modern furniture, and were ultra-colorful. They were a fun place to be waiting for your next flight. In a few words, Braniff was the Airline of Life and Color.

The following people played a major part in helping me put this book together. I owe a big thank you to Ben Cass, who is the President of the Braniff Airways Foundation. He assisted me greatly with insightful, and interesting information on the history of his great airline. He dispelled many rumors, and falsehoods about Braniff, and laid the truth of the airline on the line. I urge anyone who is interested in Braniff International Airways, to go to the Braniff Airways Foundation's website at: http://www.braniffairwaysfoundation.org/

Another person, who assisted me with many rare and interesting photographs from his personal collection, is John F. Ciesla. Not only did John, an author of many non-fiction books, offer his photos to me, he gave me a lot of assistance in putting my book together. He also gave me sound advice on the types of images that would be suitable for the book. The photographs from his collection are outstanding and have enhanced the quality of this book immensely. I consider John a good friend now, and I value his advice. In fact, we kept in constant email contact during the writing of this book. I thank him very much for everything he has done. John has a fantastic FLICKR page with some of the best photography you could hope to see. He has many rail and automobile photos there as well. Head over to his FLICKR page and take a look. Here is the address: https://www.flickr.com/photos/jfciesla/albums/with/72157634012196863

What can I say about Pat Bukiri, of the Midway Airport Historians group. He allowed me full use of any photos that I deemed to be suitable for the book. No questions asked. He is an extremely helpful, and friendly person, and I owe him a lot for his indispensable assistance. For anyone who is interested in American civil aviation from the mid-20th Century (and more recent years), go and check out Pat's great Midway Airport Historians' FLICKR page at: https://www.flickr.com/photos/twa1049g/.

The following photographers were also important in the making of the book. I must single out George Mihalek, who offered me the use of his personally photographed Braniff photos. I have to say they were all great, and it was very hard to decide which to pick. He is one of the best aviation photographers I have come across, and the quality of his work speaks volumes. I am indebted to George, for his wonderful photos which adorn this book. George also has a FLICKR page and his work can be seen here: https://www.flickr.com/photos/73886013@N06/

Trevor Ogle offered many of his own photos, and from his collection for use in this book. The beautiful orange Braniff 720 in his collection caught my eye. It was that photo which made me contact him, and he was immensely helpful with my requests for use of any of his Braniff photos. Thanks very much Trevor. You can check out Trevor's FLICKR page and look through his outstanding photographs: https://www.flickr.com/photos/100672308@N07/

I would also like to Eric 31 Ragnarok, and Maarten Visser, for the use of their photos. I greatly appreciate their help. I would like to say to Eric: Merci beaucoup mon ami. To Maarten, I would also like to say: heel erg bedankt mijn vriend. Eric has more than 45,000 photos on his FLICKR page. Go and check out his page, you will not be disappointed: https://www.flickr.com/photos/ragnarok31/

Finally, I would like to say a big thank you to Bjorn Larsson, who is the owner of the most outstanding airline timetable website on the internet. Bjorn has been operating his website since 2000, and now there are well over 50,000 timetable images on show there. Bjorn was very kind and emailed me many Braniff Airways timetable images from the airline's early years to 1982. He personally selected the best ones, and they are included in the book. Go and check out Bjorn's fabulous website. You will be amazed at the number of fascinating timetables you could find there. His website's address is: https://www.timetableimages.com/

FLEET INFORMATION:

(Highly recommended airline fleet websites which assisted in the making of this book.)

abcdlist.nl. https://abcdlist.nl/

Richard Zeman. RZJets at https://rzjets.net/operators/

Servaas C. Verbrugge. Airlinerlist at http://airlinerlist.com/

Airframes.org. at http://www.airframes.org/

Mike R. Henniger. Aerialvisuals at http://www.aerialvisuals.ca/

Joseph F. Baugher, at http://www.joebaugher.com/

SECTION 1
THE HISTORY

PRE-1948

The Beginning

Braniff, a name that epitomized the entrepreneurial spirit of the early airline pioneers, came into being in the spring of 1928. Paul and Tom Braniff, the "founding brothers" of Braniff Airlines, formed the carrier at Oklahoma at a time when many new airlines in the USA were being created regularly. The main credit for Braniff's creation should go to Paul Braniff, who was the pilot and the adventurer of the two. He had become a pilot in 1919 after spending time in the Army Air Corps as a gunnery instructor during World War 1. At war's end, he became a barnstormer, a pilot flying passenger airliners, and even flew in the National Air Tour of 1927. So, it was no surprise when, in 1928, he convinced his brother, Tom, and four other investors to launch their own airline. He also urged them to purchase a Stinson SB-1 Detroiter, which was a six-seat cabin airliner built by the Stinson Aircraft Corporation. With this aircraft, the new airline was launched under the name "Paul R. Braniff, Inc." It started operations on the 20th of June 1928, with a flight from Tulsa to Oklahoma City, using the recently purchased Detroiter. However, the airline did not last long. In the spring of 1929, it had financial pressure and was consequently acquired by the Universal Aviation Corporation, which later merged with other companies to form American Airlines.

The Thomas Elmer Braniff Era

The only thing Paul Braniff could do to rekindle his dream was to convince Tom to start all over again with the other investors. And that is exactly what they did, and this time around, they purchased two Lockheed Vega aircraft, which were the fastest planes in the U.S. in 1930. The new airline took the title of "Braniff Airways" and was incorporated on the 3rd of November 1930 at Oklahoma City. To avoid any financial issues, it was Tom Braniff, the businessman of the two, who became the President, while Paul accepted the position of Executive Vice-President of Operations. In that same month, they started flying again on their original Tulsa to Oklahoma City route. Braniff Airways also added the Oklahoma City to Wichita, Kansas route, which was extended to include Kansas City. In 1931, the route was extended further to Chicago, followed by a new route south to Dallas, which was inaugurated in 1934. And if that was not enough, they added Amarillo and Houston the following year. More routes were added over the next few years, and things were on a roll, so to speak.

Braniff also completed a merger in 1935, when it took control of Long and Harman Airlines, which had mail contracts for three cities: Dallas, Amarillo, and Brownsville. It also acquired twelve aircraft from them. The most interesting of the lot being two Ford Trimotors.

World War 2

In 1935, Braniff purchased its first twin-engine airliner, the Lockheed L-10A Electra, which brought the carriers first inflight service – paper bagged luncheons served by the copilot! Talk about multitasking. Then, in 1937, Braniff received the precursor to the great Douglas DC-3; the Douglas DC-2, which was delivered to the carrier in June 1937. The aircraft proved to be so much of a hit that Braniff went on and ordered the revolutionary DC-3 as well, which entered service in 1939. In November 1941, Braniff Airways left the Braniff brother's hometown of Oklahoma City and moved to its new base at Dallas Love Field in Texas. Previously, Braniff's maintenance had moved to Dallas in November 1934.

By 1942, after the U.S. had entered the war, Braniff Airways had five DC-2s and ten DC-3s in its fleet. Many of the DC-3s and DC-2s were placed in the service of the U.S. armed forces for the duration of the war, as were many other airlines' aircraft at the time. Just over 50 percent of the Braniff fleet was in war service by early 1942.

The remainder of the fleet did not sit idle, as an important development occurred which benefited Braniff. The Air Transport Association of America and American Airlines urged the Army Air Transport Command to adopt their plan of creating a domestic military cargo service. This was a way to earn some cash by the airlines as it would be paid for by the U.S. government. The government agreed, and the plan was implemented on April 23, 1942.

In the plan, Braniff was allocated the Southwest sector of the United States. It was obligated to do domestic cargo runs in its assigned area. The other four airlines in the agreement were American, Northeast, Eastern, and Northwest. However, Braniff was the only airline in the group that was selected by the Army Air Transport Command for an extra "non-domestic" assignment. It was required to do cargo flights from San Antonio to Panama, with flights to Mexico and Guatemala were added.

These flights nicknamed "The Banana Run" by the crews, who assisted local Banana farmers that wanted to export their produce to the United States. As it was rotting, and the farmers had no other way of getting it there, the Braniff crews helped them do it. "The Banana Run" ran until early 1944. During that time, it made about 2,300 flights with no incident. This must have been a record in a time of war, even though they were far from the frontlines. Then, on April 4, 1945, while the war was still going, Braniff inaugurated its first official civilian flight in Mexico, when Braniff's Mexican subsidiary, "Aerovias Braniff", took off from Nuevo Laredo for Mexico City.

POST WAR

Latin America

Braniff's dream for Latin America service had officially started in 1943 when it established its subsidiary, Aerovias Braniff and applied for service to Latin America and across the southern US and beyond to Europe. It did not begin operations until that flight on April 4, 1945. Oddly, the move to start its own airline in Latin America was prompted by politics. Pan American, the U.S. government's favorite airline for international route approvals, had a monopoly of international routes up to the beginning of World War 2. This situation changed when a new ruling was made that allowed other airlines to join the fray and apply for their own international flights. American Airlines was first in line, and it received approval from the Civil Aviation Board—which initially rejected its application. It was later approved by President Roosevelt after he had intervened. The new airline started flights from El Paso and Dallas to Mexico City. This was quickly followed by Braniff's application of Aerovias Braniff. With the approval of the Mexican government, Aerovias Braniff began operating within that country with a fleet of three DC-3 airliners.

After Aerovias Braniff's first flight, the carrier was also awarded the rights by the Mexican government to fly to other points in Central America and Cuba, from Mexico City. Things looked bright for the new carrier. It might have even succeeded if it was not for the anti-competitive activities of Braniff's main competitor in Latin America, Pan American, which scuttled the plans. Panagra—the Latin American arm of Pan American—was the holder of all major routes in the region, and Pan American wanted things to stay that way. Thus, in 1946, when Aerovias Braniff was still an infant, Pan American and the U.S. Government pressured the Mexican government to act. Action was indeed taken, and all Aerovias Braniff's routes were revoked on October 26, 1946. This ended Braniff's brief foray into a major competitor's back yard in Latin America. Braniff was waiting for its moment for retribution, which was to come later.

The Douglas Propliners

When the war ended, the U.S. airlines and U.S aircraft manufacturers were primed for a big entrance in the world stage. In many ways, the war had helped them immensely by giving the airlines invaluable international experience. For the manufacturers, it helped to speed up their research, development, and production of many new aircraft types. Such progress would probably have taken longer if it were not for the war. After the conflict, a plethora of ex-World War 2 aircraft was made available to the airlines at bargain-basement prices. They were spoiled for choice; Douglas C-47s and C-54s; Curtiss C-46s; and Lockheed C-69s were abundantly available. Some uncanny aircraft were also converted to airliners—Consolidated B-24 and Boeing B-17s—though these were mainly used overseas, predominantly in Europe or Latin America. Braniff was one of the airlines which took advantage of this once-in-a-lifetime fire sale. In 1946, it received its first ex-USAAF C-54s, which went on to be converted to DC-4 civil standard. This workhorse was well appreciated by the airline and quickly put into service on the Dallas to Chicago route, via Kansas City, in May 1946.

The DC-4 had only enjoyed the limelight for one year or so, when the new kid on the block arrived: the fully pressurized Douglas DC-6, of which six were ordered by Braniff. If the DC-3 had revolutionized air travel, this airliner outstripped it and put its own footprint on what air travel should be like. The DC-6 would allow Braniff to expand and fly to distant lands with comfort and ease.

The DC-6 was first constructed for the USAF as a military transport under the designation of XC-112. The USAF went cold on the aircraft, and only two were made. This was a blessing in disguise for Douglas as they went on and produced one of the greatest civil airliners in its time. In fact, American Airlines put in the first order for the new prop liner late in 1944, followed closely by United Airlines. TWA went with the Lockheed L-049 Constellation. There was a major rivalry going on between American, TWA, and United—the big three intercontinental carriers. Braniff, Eastern Airlines, and Delta were in the upper tier, but not being intercontinental airlines, they went along doing their business seamlessly. The first DC-6s were delivered to American Airlines and United Airlines in November 1946. Braniff was not too far behind, it received its first DC-6, registered N90881, on August 23, 1947.

Meanwhile, back in early 1946, Braniff's top men, Tom Braniff, and his protégé, Charles Beard, were planning their future expansion into Latin America, on the expectation that the Civil Aeronautics Board was going to give full operational rights to Braniff for that market. They planned every move they wanted to make methodically, astutely, and with great foresight. That was wise, as finally, on May 19, 1946, Braniff was awarded those rights by the CAB. It must be said; President Truman had a hand to play in the decision when he gave his full approval of Braniff's application. The reason being, he was not too happy about Pan American's tight control of the Latin American market.

POST-1948

The Birth of Braniff International Airways

Braniff finally got what it wanted. The Latin American region was there for the taking. To prepare for the situation, Braniff ordered six new DC-6 airliners after the approval by the CAB. Now, with the DC-4s, which incidentally had only a handful of years of service left in them, and the new DC-6s at its disposal, Braniff was finally ready for its first scheduled international service to Latin America. Thus, when the shiny new Braniff Airways, Douglas DC-6, took off from Houston, Texas for Guayaquil, Ecuador, with stops in Cuba and Panama, on June 4, 1948, it was the beginning of a new era. Also, service was extended to Lima, Peru, a few days later. Of course, on the very day that the DC-6 was making history, the company was so delighted the name of the carrier was changed to "Braniff International Airways." The "International" trade name was actually adopted on May 19, 1946, the day when the new routes were awarded. Latin America was ready for the airline with humble beginnings in Oklahoma, but with great International dreams.

Braniff's eventual entry into Latin America was three years in the making since the ill-fated adventure with Aerovias Braniff. Then, Pan American and its confidant Panagra, took all the awards. Now, however, with Braniff back and barking at their heels, the heat was on both of them.

Panagra, the South American-based airline formed by Pan American World Airways and Grace Shipping Company, must have been more than a little worried by February 1950. By that time, Braniff had a growing international network in Latin America with stops at Mexico City, Havana, Panama City, Guayaquil, Quito, Bogota, Lima, La Paz, and Rio De Janeiro. In March, it expanded further when it opened a new service to Asuncion, Paraguay, and then in May to Argentina's capital, Buenos Aires (literally "Fair Wind"). Not to upset the Brazilians in the North, Braniff opened a service to Sao Paulo in October 1951.

The Mid-Continent Airlines Takeover

The Braniff domestic network was not neglected either. By mid-1952, Braniff had created a solid domestic network from the Mid-Western states all the way to Denver, Colorado. To top it all off, in August 1952, it acquired one of the better smaller trunk airlines in Mid-Continent Airlines (previously known as "Hanford's Tri-State Airlines"). This airline flew to twelve states and thirty-five cities; from Houston, St. Louis, and New Orleans, to places further north, such as Minneapolis/St. Paul, Sioux Falls South Dakota, and Minot North Dakota, right next to Canada. It was quite a busy outfit.

The acquisition of Mid-Continent Airlines, however, did not enhance Braniff's fleet too much. To be sure, it received 23 Douglas DC-3s, which came in handy, but were too slow and could not compete with the newer and faster pistons of the 1950s. Braniff also acquired four Convair 240s in the Mid-Continent Airlines merger. But these were utilized for one year before being withdrawn from service in 1953 and sold to the Italian national airline, Linee Aeree Italiane. Braniff's own DC-4s, which had served the airline since the mid-nineteen forties, did not fare much better. They were also withdrawn from use and were sold by November 1954, except for the one DC-4, which was destroyed in a non-fatal accident in March 1952.

Braniff DC-3 N25672 at Chicago Midway Airport in 1953. *Photo courtesy of Pat Bukiri and the Midway Airport Historians.*

Braniff was bent on improving its overall service and efficiency. The acquisition of new aircraft was necessary to offer a better and faster service for its domestic customers, which meant that the older aircraft had to be replaced. With that in mind, Braniff made a massive order for 35 new Convair CV-340s, and all were delivered between July 26, 1952, and October 28, 1953. Here was an aircraft, which at the time was at the cutting-edge of technology. Its predecessor, the CV-240, was too short at 74 feet 8 inches and could only carry 40 passengers. With the Convair CV-340, however, Braniff got more bang for its buck. It was 79 feet 2 inches in length, allowing for the carrying of 44 passengers, and had an increased wing area, which meant it allowed for a higher gross weight and better performance at high altitude. This aircraft was a game-changer for Braniff, and it eventually replaced the carrier's fleet of DC-3s.

THE CHARLES EDMUND BEARD ERA

They say every cloud has a silver lining. In this case, the cloud was a dark and ominous one that rattled the company in January 1954. The founder, CEO, and inspirational leader of Braniff, Thomas E. Braniff, died with ten others in a private aircraft accident while on a fishing trip at Louisiana on January 10. The company, with the person whose very name was adorned on every one of its airliners was gone. Who could stand up and be counted on in this time of need? The answer to that question arrived three days after the accident when Charles Edmund Beard was elected President of Braniff International Airways.

Sadly, another turn of fate befell Braniff when Paul R. Braniff died after an eight-year battle with bone cancer on June 1, 1954, just five months after his brother and followed by the untimely loss of Mrs Thomas Elmer Braniff from brain cancer in August 1954. The Braniff brothers were both gone, but their airline was in good hands under the new President. He was responsible, honest, approachable, and a genuinely likable individual. He would play an important role during this turbulent period by setting the company on a more stable footing.

Some had suggested that Charles E. Beard was too conservative for an airline such as Braniff. Many disagree with this opinion. If there was one person who could be attributed to Braniff gaining profits, increasing traffic, and making judicious decisions, it was its new President, Charles Beard. Let us not forget that it was he who led Braniff into the jet age. He was a visionary. He placed orders for five 707-227s in September 1955, at a time when many piston liner devotees in the airlines derided jet-liners as being too costly and inefficient. They preferred prop liners because they thought they were much cheaper to run and were more reliable. In their minds, the good old DC-6s and Constellations were heaven-sent, unlike the evil jets. His decision to pull Braniff into the jet era that early in the game was pure genius.

The quest for further improvement was unquestionable as new orders for aircraft kept coming from Braniff. In 1955, the carrier put in orders for 7 DC-7Cs, 5 Convair CV-440s, and the Boeing 707-227s. Interestingly, before these aircraft were delivered, Braniff made two purchases of used equipment to enhance its fleet. The first purchase was of two Lockheed L-049s from Linea Aeropostal Venezolana (LAV), for domestic flights to four states: Texas, Oklahoma, Kansas, and Illinois. Both were delivered in August 1955. They served their term admirably and were both sold by March 1961 to Hubert M Covert. The second was for two Curtiss C-46s which were purchased from Civil Air Transport in November 1955. They were used as freighters to supplement Braniff's cargo services. They were eventually both sold in October 1963 to Air America. Tragically, one of the two crashed on takeoff after an engine failure at Laos on November 25, 1968, with the loss of eleven lives.

Later, in February 1956, Braniff ordered even more brand-new aircraft, when it placed an order for 9 Lockheed L-188 Electras, the fastest turboprop in the world at the time. Incidentally, deliveries of the new DC-7Cs started in September 1956 and were quickly put on the Dallas to New York flights in November. The last DC-7C was delivered to the carrier in May 1957. As for the Convair CV-440s, five were delivered to Braniff between November 1956 and January 1957. They were placed on local service routes and were quite successful.

Braniff International reached a milestone in 1958 when it became the first airline in the world to introduce a computerized reservation system. This, together with the opening of a new terminal at Love Field, the move to the new headquarters at Exchange Park near Dallas Love Field, and the new Braniff International

Airways Maintenance and Operations Base at Dallas Love Field, which took two years to complete, showed the improvement made by the Texas carrier under the leadership of Charles Beard.

The Jet Age

1959 was the year that propelled Braniff into the jet age. That was the year that it took delivery of its first turboprop, the Lockheed L-188, on April 29, 1959. This was a sure sign to the competition that Braniff meant business and was not going to be a second-tier operator with only piston propliners. Yet there was more to come; the 707s were also on their way. In those years, an airline receiving the new Boeing 707s was like receiving a Space Shuttle. So much was the hype back then with the new jetliners, not only in the U.S. but also around the world. In the four years from the placement of the orders in 1955 to the actual arrival of them in 1959, Braniff's President, Charles Edmund Beard, knew that going with the new jet-powered transport was always going to be the right move. The age of the jets was here; there was no going back.

The five 707-227s were based on the 707-120 type, which was powered by Pratt & Whitney JT3C-6 turbojets. Pan American flew the first one after FAA certification in October 1958 from New York to Paris. That was Pan Am's first scheduled jet flight. As for Braniff, the 707-220 type was more up its alley. It was an improvement on the 707-120 by way of takeoff performance and was fitted with Pratt & Whitney JT4A-3 engines with 15,800 pounds of thrust. It was also the fastest jetliner in the world. Interestingly, only five 707-220s were produced, and four were delivered to Braniff, while the fifth one was destroyed during a Boeing test flight crash one day before delivery. Braniff received the first of the 707-227s, N7072, on December 3, 1959. Just sixteen days after delivery, the first-ever flight of a pure jet liner by Braniff occurred on December 19, when the new jetliner in the Eldorado scheme took off from Dallas for New York.

Braniff BAC 1-11 at Houston-Hobby airport in 1969. "Braniff International BAC 1-11". By Clipperarctic. https://www.flickr.com/photos/23344035@N03/6154295000/in/photolist-bBbAe9-anQnEf-a9BiEA-cq1PY1-ak2P1Z-a5mAtR-anQnuu is licensed under CC BY 4.0 https://creativecommons.org/licenses/by/4.0/. Cropped from original.

The DC-6s and DC-7Cs might have helped Braniff to expand to South America more than it could ever do with the older DC-3s and even DC-4s. Then, with the introduction of the 707s, Braniff initiated its first jet-powered South American service. As for domestic and Mexico services, Braniff used the new turboprop L-188s. It is sufficient to say that the piston liners were blown out of the water with the sheer speed, grace, smoothness, and efficiency of the new "Jet" powered airliners. The era of the pistons was quickly coming to an end.

With the acquisition of the jets, Braniff's route system by February 1, 1960, looked quite complex, with so many new destinations. It covered the length and breadth of Latin America. There were the Central American destinations of Mexico City, Panama City. There were also destinations to South America, including Bogota, Lima, La Paz, Rio De Janeiro, Sao Paulo, Buenos Aires, Santiago, and Asuncion. It had Latin America covered.

The 707s got to enjoy some time with their little brothers, when three new Boeing 720-027s joined the Braniff fleet in 1961. They were nicknamed "El Dorado Super Jets," as were the 707-227s. A further two were delivered in May 1962 and August 1963. They were mainly placed on domestic routes to Minneapolis/St. Paul, Kansas City, Dallas, and San Antonio. The only exception was a route to Mexico City.

Braniff was on the lookout for a short-range jetliner for its regional routes to replace the Convair 340s & 440s, as well as DC-6s and DC-7s on domestic routes. Beard decided to place an order for 6 British BAC-111-203AEs, with an option of 6 more. The British jetliner was not expected to make its first flight until August 20, 1963. Even before the first flight, Braniff took up its six options on March 5, 1963, with another two BAC-1-11s ordered on February 14, 1964—Valentine's Day. The first of 14 BAC-1-11 airliners arrived at Dallas on March 11, 1965, and all were delivered by December 8, 1965. Braniff created another record when, in April 1965, its inaugural flight from Corpus Christi to Minneapolis with the BAC-1-11 was the first short-haul flight with a pure jetliner in American aviation history.

The Panagra Purchase

As mentioned earlier, Pan American and Panagra (Pan American Grace Airways) must have been looking askance at Braniff's intrusion onto their South American turf. In fact, in the early 1950s, Beard had suggested merging with Panagra but Pan American vehemently buried any talk of a takeover of Panagra. Then, in April 1963, the Panagra issue resurfaced again when Beard and Braniff initiated talks between themselves and Panagra to convince them to accept Braniff's offer. However, they failed to see eye to eye and everything was back to square one.

The dream would not die for Braniff. It wanted the Panagra merger at all costs. The negotiations reached an impasse. Something had to give, and that something happened in 1964 when Braniff and Panagra continued their next round in the negotiations. This time, President Beard negotiated a 22 million USD agreement whereby Braniff would buy W.R. Grace's half of Panagra for 11 million USD, and purchase Pan American's half of Panagra for another 11 million USD. It was also agreed that the merger should take place by early 1965. Panagra liked what it heard and accepted the offer. Braniff would now have to convince Pan American to accept its part of the offer. Braniff was heading for negotiations with the master tactician, the iconic President of Pan American, Juan Trippe, who was a wily and seasoned negotiator. The meeting went ahead, and things were cordial, but Trippe, who almost certainly knew what Braniff was going to offer, gave his answer. Unfortunately, it was not good news for Braniff. Trippe shut down any talks of a buyout of Pan American's shares of Panagra. Negotiations had reached a stalemate once again. For Beard, it was a huge disappointment. The merger he had wanted for Braniff looked as if it was not going to materialize during his watch. Then something did happen.

As far back as 1953, the Civil Aeronautics Board had requested the Department of Justice to file charges against Pan American for violations of the anti-trust laws by limiting Panagra, not allowing the carrier to extend its routes northward into the U.S. and Canada. Central America was as far as Pan American allowed

Panagra to fly to. On May 8, 1961, a U.S. District Court for the Southern District of New York handed down a decision stating that Pan American violated the "Sherman Anti-Trust Act" by limiting the expansion to Northern America for Panagra. The court also ruled that Panagra was cleared of any wrongdoing and that the only one that was in violation was Pan American. Because of that, it should divest itself of Panagra stock and its fifty percent ownership of the airline. Then, on January 14, 1963, the case was decided at the Supreme Court, which ruled that the case fell under the jurisdiction of the Civil Aeronautics Board, which essentially meant that the CAB could now force Pan American to divest itself of Panagra completely. The decision by the Supreme Court sent raptures throughout Braniff. Their major rival in the region was now going to be a part of Braniff.

In December 1965, Braniff International successfully negotiated the purchase of W. R. Grace and Company's 50-percent share in Panagra. However, the total amount had now been negotiated to 15 million USD. Then, in March 1965, Pan American also negotiated for the same amount. The total was 8 million more than Braniff had originally offered, but it was happy with the result. It looked like Braniff paid much more than they anticipated, in reality, Braniff came out ahead as five new Panagra DC-8-62s on order, were already financed and paid for and Braniff netted that additional $27 million in the merger all thanks to Harding Lawrence's brilliance in this negotiation with Trippe. Trippe later said that Harding Lawrence was the best airline manager he had ever come across and was amenable to a Braniff Pan Am merger that was proposed in 1970. The official merger of Braniff and Panagra was complete on February 1, 1967.

The saga was finally over, but there was one major casualty: the much-liked President of Braniff International Charles E. Beard was gone. During 1964, while the drama with the Panagra deal was still going on, an agreement was made to sell Braniff International Airways to the founders of Texas Instruments, going by the name GreatAmerica Corporation, which was controlled by the insurance magnate, Troy V. Post. He had big plans for Braniff and one of them was the replacement of Charles Beard with Harding Lawrence, the ex-Continental Airlines Executive Vice President since 1958. Harding Lawrence was one sharp cookie. He was mentored by the highly intelligent and driven President of Continental Airlines, Robert Six, one of the great pioneers and visionaries of the U.S. aviation industry. While Lawrence was at Continental, he presided over a 500 percent increase in sales during his time there.

Charles Beard was a good President, but he was not in the same league as people like Six and Lawrence. Lawrence was more business savvy, razor-sharp, and a tough man who was hard to beat. He did have scruples and cared about Braniff and its employees – he coined the term "The Braniff Family." He eliminated the company's nepotism policy that allowed several generations of Braniff Family members to work for the company at the same time. Concerning Beard, he was liked by all the employees at Braniff as he was relaxed, friendly, and approachable. Still, he was not forceful, and with the stiff competition from rival airlines coming in from all angles in the new jet era, there was no hiding. The new breed of airline executives had to have nerves of steel, astuteness, and be able to make hard decisions, Lawrence was the man to steer the ship. The replacement of Charles Beard was always going to happen, especially at an airline like Braniff, which was one of the top ten U.S. airlines. The employees of Braniff were sad to see their amiable President leave them, and waited to see what was waiting for them in the wings with the new administration.

Amidst the Panagra saga and the takeover of Braniff by the GreatAmerica Corporation, Braniff took delivery of the following aircraft types: a used 720-048 from El Al airlines of Israel on September 15, 1964. They leased two others, one from Aer Lingus of Ireland in January 1965, and one from EL AL in November 1965. The thirteen BAC 1-11 deliveries were also made.

THE HARDING LAWRENCE ERA

The End of the Plain Plane

After the dramas of the past two years—with the Panagra saga and the takeover of Braniff-Lawrence's goal was to change Braniff's image completely. He thought the airline lacked life, color, vibrance, and pizzazz. Who could blame him? Lawrence wanted colorful aircraft, new logos, new catchphrases, new groovy advertisements, and colorful, sensuous outfits for the air hostesses. He might have been a great aviation administrator, but he was no fashion or advertising guru. He hired one of the best advertising executives going around in Mary Wells. He knew Wells very well as they had both worked on an advertisement campaign while he was at Continental Airlines. Of course, their reunion was quite convenient, and they got on so well this time around that they got married in November 1967. They were also the first power couple of New York.

Wells changed the way Braniff was viewed with her new and transforming marketing ideas. She created the "The End of the Plain Plane" advertisements which were risqué in the mid-sixties, where conformism and social mediocrity were still the norm. This was two years before the flower people and the summer of love changed America forever. Wells' ads had color, frivolity, and sensual desire, all coming out from the TV screens. Who could not feel a little uncomfortable while looking at the "Air Strip" advertisements? The ads did work though, and Braniff was seen as a very progressive airline compared to the likes of Pan American, TWA, United, or American. She also hired one of the top designers of the era, the Italian master of color, Emilio Pucci, to design the colorful uniforms for the air hostesses. Let us not forget the iconic 1965 Solid Color Scheme, which was coined the "Jellybean" scheme by the press. For that, Wells hired another famed designer and architect, Alexander Girard, who insisted the planes would look more recognizable from the ground with the proposed new colors. The aircraft fuselages were painted in seven different colors. Girard also wanted all the aircraft bodies, tails, wings, and engines in one color. Braniff did not agree with this proposal and decided to have the tail, wings, and engines painted in white, which was how the Braniff Lockheed Vegas' were painted in the 1930s. Aside from the "Jellybean" livery, Girard put colors on everything—furniture, ticket offices, customer lounges... you name it. Braniff International was full of color alright. That never changed until the cessation of operations in 1982.

Boeing's Airline

In 1965, Lawrence ordered 12 727-100 airliners and canceled the rest of the BAC 1-11 orders. The 727s had been selected before Lawrence took over, but no orders were made. They were the QC (Quick Change) models, where the passenger seats were placed on pallets rather than bolted on the floor, which allowed Braniff to take out all the seating and set up the aircraft for freight duties within 30 minutes. The aircraft also had a large freight loading door on the left front side of the fuselage. The 727QC enabled Braniff to increase its profits by using the QCs to fly passengers in the daytime while ferrying freight overnight. This was a smart move by the airline as the aircraft were used efficiently. Braniff was so pleased with their efforts that it started its own cargo business named 'AirGo,' and its cargo containers were called Cloudpacks.

The 727s were, at the time, extremely popular with many airlines throughout the U.S. Boeing was selling airliners in the early sixties like hotcakes. It was a time when Douglas started to play second fiddle to

Boeing. To be sure, Douglas did well in sales and produced some beautiful airliners, but the swing towards Boeing airliners was obvious. Boeing was mainly a major military aircraft manufacturer and it had been only a small airliner manufacturer only ten years before. The 707 changed all that. Douglas must have been a little worried when Boeing announced the production go-ahead of their first jetliner in 1955, and the historic order for 20 707s by the number one international airline in the world—Pan American. From this point on, Douglas would be chasing Boeing's tail. Boeing has always set the pace since that first 707 flight. The new Boeing jetliner came out a year before the DC-8 and was an instant success, thanks in no small part to Pan American. It was in the spotlight at all the destinations that Pan American flew to.

Although many airlines did not buy Boeing but waited for Douglas to release the DC-8, such as Delta, Air Canada, KLM, Alitalia, and Iberia, most did not wait. They bought the new Boeing super-jet and were satisfied. To rub salt into the wound: the 727 program was announced by Boeing on December 5, 1960, with Eastern and United ordering 40 aircraft each. Most importantly for Boeing, all their 707, 727, 737, and even the 757 aircraft had the same fuselage diameter, which allowed for a more time-efficient and smoother production process. They could roll out jetliner after jetliner like it was nothing. Over the years, the Douglas/McDonnell Douglas company tried valiantly to get ahead of Boeing, but the latter company was not going to give an inch. The end was near for McDonnell Douglas. One fateful day in August 1997, Douglas was subsumed into Boeing. The illustrious company, which based most of its history in producing so many great airliners, was forever gone.

Braniff International Boeing 727-027C N7279 "Orange Jellybean" at Chicago O'Hare Airport in June 1967. *Photo courtesy of the John F Ciesla Collection.*

In the 1940s and 1950s, Braniff was a minor Douglas operator compared to the major transcontinentals. It did have many DC-3s from the Mid-Continent Airlines takeover. Its DC-4 fleet stood at ten aircraft, the DC-6 fleet at eleven aircraft, and the DC-7 fleet at nine aircraft. Although Braniff did operate the 707-227s before Lawrence became President, once he took over the reins in 1965, Braniff became a major Boeing purchaser as the majority of Braniff's orders under his leadership were of Boeing jetliners. From the first 727 delivery in 1966 until its demise sixteen years later, Braniff would operate a total of one-hundred and thirty-one 727 airliners. Its fleet also consisted of seventeen 707s (four-227s; nine-327Cs; four-138Bs), nine 720s, and eleven 747s. Braniff did have a few DC-8s in its fleet until the close down in 1982, but the fleet was predominantly made up of Boeing aircraft.

Braniff might not have ever had Douglas jetliners in its fleet if it was not for the merger with Panagra in February 1967. In the takeover, Braniff adopted a fleet of four DC-8-31s, 1 DC-8-55F, along with prepaid deposits of $27 million USD on five ordered DC-8-62s, which were delivered to Braniff in the summer of 1967. However, the DC-8-31s and DC-8-55F were not needed. By November 1967, all four DC-8-31s were sold off. One was bought by United Airlines, and the other three by Capitol International Airways, one of which crashed during a training flight on April 28, 1968, luckily with no fatalities. The solitary DC-8-55F was also sold to Canadian Pacific Airlines on November 17, 1967. Later, Braniff did purchase more DC-8s as it decided they were suitable for its needs. Overall, a total of twenty-one DC-8 types went into Braniff service: DC-8-31, DC-8-51, DC-8-51F DC-8-55F, DC-8-62, and the DC-8-62CF.

The All Jet Era

In 1966, Braniff scored a major coup. Its owner, Troy V. Post, sweet-talked his way into a good deal for the carrier. He was on good terms with the U.S. President and he managed to get a government contract to transport military personnel in the Pacific region. These flights flew the soldiers from South Vietnam to Hawaii for their R&R furloughs during the Vietnam War, for the Military Airlift Command (MAC), previously known as the Military Air Transport Service (MATS). This was later supplemented to include transatlantic flights to Nova Scotia, Greenland, Iceland, and Germany for MAC. These flights lasted until 1968. The Pacific deal was also expanded to include services to Wake Island, Guam, Kimpo, Kadena, and Yokota bases in Japan, Clark AFB in the Philippines, Saigon, Cam Ranh Bay, Danang, and Bien Hoa in South Vietnam, Vientiane in Laos, and Bangkok in Thailand. The flights for the Pacific region originated from Travis AFB in California and Braniff used its newly received 707-327Cs for this service. A total of 9 707-327Cs were used and they had been delivered to the carrier between May and November 1966. For the Atlantic region, Braniff used the Boeing 727-27QC aircraft, and all flights originated from McGuire AFB in New Jersey. The military flights ended in March 1973, just as the Vietnam war was starting to wind down.

1967 was a fateful year for the prop liners of Braniff International. That was the year when all propeller-powered aircraft were phased out. All DC-7Cs were pulled from scheduled service in 1966; the Convairs were pulled in September 1967 with the arrival of the first DC-8-62; and all DC-6s were pulled in late 1965. Braniff International's fleet was now a fully jet-powered one. While the prop liners had gone, 22 jetliners took their place. The actual number was 27 deliveries, but four of those were the DC-8-31s and one DC-8-55F from the Panagra takeover, which were hardly used by Braniff and sold in late 1967. In 1968, the actual deliveries of jet aircraft that joined the Braniff fleet were as follows: thirteen 727-27QCs; all bought new from Boeing, 4 DC-8-62s and one 62CF; and four 707-327Cs; all also bought new from Boeing. Braniff was happy with the DC-8-62s, so it ordered another two models from Douglas, which were delivered to the carrier in October and November 1969.

"Braniff International 707-327C departing from Phu Cat Air Base in 1971." By Al Walker via Tommy Japan 79. https://www.flickr.com/photos/97930879@N02/10490631805/in/photolist-gZ2ema-pc7h5wis is licensed under CC BY 4.0. https://creativecommons.org/licenses/by/4.0/ . Added saturation, and sharpened image.

You'll Like Flying Braniff Style

It must be said that over the years, Lawrence made large profits for Braniff; he also achieved record revenues and traffic numbers. He was always ahead in terms of keeping the airline moving upwards, and advertising was his vehicle to achieve those ends. In 1968, Lawrence was on the lookout for an agency to take Braniff once again to another level, three years after the "The End of the Plain Plane" advertising campaign. He could not use Mary Well's agency as Mrs. Wells was offered the TWA account. As she could not have two airline accounts, she took the TWA account and left Braniff on December 1, 1968, but she did come back in 1979. Thus, Lawrence had to look elsewhere. The big agencies—and there were many of them—knew about this and were eager to please the man from Dallas. Lawrence narrowed his list of agencies to five and decided to go with the agency run by George Lois of New York, Madison Avenue. Lawrence thought they had the best slogan of all, which was: "When You Got It, Flaunt It." It was a great advertising campaign, and many in the advertising industry saw it as one of the best campaigns they had ever seen. The campaign doubled Braniff's business, but it was seen as a failure by the press. The news agencies must have had it in for Braniff (or Lawrence), as they claimed the "When You Got It, Flaunt It" campaign was an unmitigated disaster for Braniff. What they failed to report was that America was in recession 1969 to 1970, which was the main reason why losses were incurred by Braniff. As for the campaign itself, the Lois "When You Got It, Flaunt It" campaign was one of the best campaigns ever made for an airline company. In the TV advertisements, it listed every major star in the United States at the time, from Andy Warhol to Sonny Liston. Braniff International did make its first minor loss of $2.6 million in 1970 under Mr. Lawrence However, over $2.3 million in struck airline payments made up most of the loss. In comparison to other carrier's losses during this time, this was modest.

Braniff had to create a new campaign to dispel the disparaging reports in the press. The next campaign had to be universally successful, especially as seen by the press. If there were going to be another criticism of Braniff's campaign, it would lose more market share in the cutthroat airline industry. Luckily, the next ad campaign did ease Braniff's customers' worries as it once again refocused on the customer. It was the "You'll Like Flying Braniff Style" campaign and the brainchild of Clinton E. Frank, who scored his biggest touchdown of all with that. The jingly tunes they created for the campaign were a little quirky, but they did the job. The customers did come back, but the end of the recession also helped improve passenger numbers.

In late 1969, Lawrence, once again, got Clinton E. Frank on a task to create a new mini advertising campaign for the brand-new Boeing 747, which was going to be delivered to the company on January 5, 1971. It did not take long for Frank to come up with the "747 Braniff Place" and "The Most Exclusive Address in the Sky". For this campaign, the 747 was going to be like you were staying in one of the most exclusive residences in the world, rather than a large metal object flying high over Texan skies, spewing black smoke out of its JT9D-7 engines. The slogan did catch on, and the "747 Braniff Place", or the Great Pumpkin by employees, became a favorite for Braniff's customers. It must be said that many aviation enthusiasts nicknamed the 747 "The Great Pumpkin." Pumpkin or not, she was a sight to behold when flying overhead. Braniff created a great livery for the airliner and she flew for the airline until the end.

The fleet had a major change when all L-188 Electras were removed from service in April 1969. They were the last propeller-driven aircraft to fly in Braniff colors. With the Electras on their way out, Braniff decided to purchase four used 707-138B airliners from the Australian airline, QANTAS, and they were delivered to the carrier between June and September 1969. QANTAS is always a good airline to buy used aircraft from. They have one of the most stringent safety regimes in the world, and the aircraft are always serviced on time. QANTAS has the best safety record in the world and has never had a hull loss since 1952. Braniff kept the 707-138Bs until all four were removed from service in 1973 and replaced by DC-8-51s. One of Braniff's 707-138Bs, registration number: N108BN, was sold to Tracey Investment in September 1975. It was eventually sold to John Travolta on May 20, 1998, and he flew it around the world in the classic 1960s colors of QANTAS. The deal he had with QANTAS was a good one for him. QANTAS would maintain it while he owned it—not a bad deal at all.

The Braniff fleet in 1970, before the delivery of the 747, consisted of the following types: four 707-227, four 707-138B, nine 707-327C, five 720-027, fifteen 727-27, eighteen 727-27QC, seven DC-8-62, and thirteen BAC 1-11 aircraft. It had on order three 747-127s, although only one was delivered: N601BN.

Although the fleet looked healthy in 1970, for Braniff and the airline industry as a whole, the year was a torrid one. The reason was the recession which lasted almost one year from December 1969 to November 1970. The recession was the worst since the Great Depression, and it greatly affected the airline industry. It was the worst financial year for U.S. airlines in history. Huge losses were made across the board. It was the perfect storm. Most airlines had brought in more capacity, and most of that capacity was taken up by the Boeing 747. With the recession, anything the airlines bought skyrocketed up in price, much faster than the inflation rate. Passenger numbers were also lower, even though the cost of a ticket had dropped steadily since 1960. Worst of all, in a normal year, wages take up about 50% of the airlines' costs, but with airline wages also having gone up by 15% on average in 1969, it was a bone cruncher. The CAB had to intervene, and they did so by approving some small fare increases, but they were too little too late as the airlines' finances were spiraling downward faster than a Himalayan avalanche. The airlines had their backs to the wall. The only remedy, if you can call it that, was to cut costs by limiting services to a minimum, and that, of course, led to many job cuts. In total, 12,000 employees were on furlough by the end of 1970. Regarding Braniff, it fared the best of all U.S. carriers. There was a very small loss of $2.64 million, with $2.3 going to required payments to struck airlines. Braniff retracted, and shrewdly only bought one 747 Jumbo Jet instead of three, and no DC-10s or L-1011s, thereby avoiding the overcapacity that killed the airline industry at the time.

1971 saw the end of service for Braniff's unique 707-227s. They were traded to British West Indies Airways (BWIA) of Trinidad and Tobago between February and May 1971, for 3 of BWIA's 727-78s, which were sent to Braniff between March and June 1971. The 727s that Braniff received were not in a satisfactory condition and had to be overhauled. The engineers at Braniff did their tasks professionally and had them ready for service in good time. Interestingly, one of the 727-78s that Braniff bought from BWIA registered N306BN, became the 727, which holds the record with the highest number of landings, with over 79,000 landings. That record was achieved in December 1999 while in service with the Bolivian carrier, LAB.

Southwest Airlines

The next chapter in Braniff's history was a "civil war" between itself and the two other established Texan companies, and a small Texan airline with a name that many people know today as Southwest Airlines. The war was bitterly fought. There was no space for calmness, amity, and reason now. The battle lines were drawn and only one side could walk through the ring of fire into victory.

Southwest Airlines was originally formed on March 15, 1967 under the name of "Air Southwest" by a lawyer named Herb Kelleher, and his partner, investment consultant, Rollin King. Their plan was simple. They would base their airline in Texas and initially offer a low-fare service in the Texas triangle, which contained five of the 20 biggest cities in the US, and was home to more than 70% of all Texans. The cities in the triangle are Houston, Dallas-Fort-Worth, San Antonio, and Austin. Their airline was mainly based on the Californian carrier's "Pacific Southwest Airlines" (PSA) model, which could be described in one word: economical. The news rang alarm bells in the boardrooms and hallways of all three established Texan carriers—Braniff International Airways, Continental Airlines, and Trans-Texas Airways, which later became Texas International Airlines.

After the application by "Air Southwest" was approved on February 20, 1968, all three airlines started legal proceedings to stop the upstart dead in its tracks. The battle was fierce, the costs were high, and the blood was boiling. Southwest fought tooth and nail to get accepted in a state which claimed to be one of the freest states in all of the USA. After four years of tempestuous court battles culminating in victories in the Supreme Court of the United States in December 1970, and the Supreme Court of Texas in June 1971, the winner emerged, tired, worn-out, and bloodied, but with the fist of victory clinched in the air. It was not any of the three giants that were the winners, but the mouse that roared: Air Southwest. Fundamentally, this situation would probably not have happened today, as many pundits at the time felt that Southwest's premises were weak at best, but they somehow got in the ear of the courts and won. Whatever the case, when the dust had settled, they changed the airline's name to Southwest Airlines. Their first flights took off on June 18, 1971. The feud between Braniff and Southwest, which were both based at Dallas Love Field, continued until 1974. That year, Braniff moved to its new base at Dallas Fort Worth Regional Airport. Southwest from then on grew at a fast pace, and today it is the world's biggest low-cost airline, with well over 700 aircraft in its fleet.

Recession

Braniff and Lawrence must have been unhappy with the victory of Southwest. First, they had to deal with the drop in passenger numbers, caused by the recession which had ended just the month before in November 1970, and now in December, they had a dangerous competitor at their heels, threatening to cut further into the profits of the carrier.

At least Braniff could claim one victory out of the recent upsets. Many U.S. airlines that had purchased new 747s and expected the money to be rolling in were very wrong. The 747s were purchased on the belief that they would bring the cost per passenger down. But because of the recession and the CAB not offering any new routes to the carriers, the 747s were mostly half-full at best. So, if Continental Airlines, which had purchased a few too many 747s, was flying a long-haul route with only 30% of the seats occupied, they would be running the costs per passenger sky-high. The same would be the case if they would be flying on short uneconomical destinations. So, from hero to villain, the 747 was one of the reasons why many airlines had made losses in the frenzied years of the early 1970s. Many airlines were lining up to have the 747 in their fleet, more as a trophy than an actual plane that they needed. Continental Airlines did have that problem, and only three years after buying a bunch of 747s, they condemned three of them to the desert in New Mexico, letting them go to waste. That was a better choice than keeping them flying as they would have pushed the carrier to its financial limits. Braniff was one of the lucky ones. It purchased only one 747-127 and utilized it on the Hawaii run, which proved very profitable for the company as the Jumbo Jet usually had most of its seats filled.

From 1971 to 1973, Braniff International made a good recovery as it posted increasing load factors. In 1971, the load factor was 44.6%. 1972 saw a sharp rise to 49.4%, and in 1973, it went up slightly to 50.1%. That was sweet music to Lawrence's ears.

Flying Colors

What better way to celebrate the good news than to improve its image? Braniff stayed one step ahead of the competition when it introduced its new livery in 1972 named simply as "Flying Colors." In true Lawrence fashion, they were very colorful and two-toned. There were only four livery paint schemes made available, compared to the 7 for the "Jellybean" liveries, but the two tones made up for that. The new scheme with its unique two-tone color contrasting looked so much more eye-catching, vibrant, and modern. The colors used were Green/Olive Green, Red/Tan Aztec Gold, Blue/Light Blue, and Orange/Mustard Ochre. Braniff International was always ahead of the curve when it came to image. Was there any other airline in the USA at the time that could match Braniff in that game?

Also, between 1971 and 1973, Braniff made great changes in its fleet structure by selling most of its older aircraft. This was due to the carrier's "Fleet Standardization Plan" that called for only three aircraft types 727s, DC-8s and 747s. All of its 707-327C aircraft were sold. Four were sold outright—one to TMA of Lebanon, one to Varig, and two to Malaysia-Singapore Airlines. The other five were initially leased before eventually being sold to the lessees. During the same period, all the BAC 1-11s were sold by the end of

1972, except for N1543, which was sold in 1977. Three of its 720s were disposed of in the mid-sixties, of the six remaining ones, four were sold by late 1973, one was a leased 720 from United Airlines and was returned in December 1972, and the last one survived until 1975 due to it being leased by Aeroamerica until June of that year, when it was also disposed of.

At the same time as the cleanout was going on, Braniff was injecting the fleet with a set of new and a few used 727-200s to replace the old departing jetliners. Braniff did have four 727-200s which were with the carrier for 3 to 4 years. With the new group of twenty-four 727s, Braniff now had a total of twenty-eight. This was by no way the final number, as by June 1980, another fifty 727-200s would join the fleet, making it a total of seventy-eight of the 200 model. In comparison, arguably, the biggest airline in Western Europe—German carrier, Lufthansa—only had an all-time total of thirty 727-200s in its fleet. Braniff was one of the biggest users of 727-200s in the world.

Alexander Calder

Things looked like they were settling down. Harding Lawrence, however, had other plans on the horizon. He was never short of highly effective and profit-making ideas since he became the chief at Braniff, what with the "Jellybean" livery, the remodeling of the outfits for the hostesses so they could look more sensuous, and the "When you've got it, flaunt it" campaign, he once again surprised everyone. In 1973, he commissioned the famous sculptor from Pennsylvania, Alexander Calder, to paint a DC-8-51 in the "Flying Colors of South America" scheme. The scheme was created to celebrate 25 years of service to South America. To be sure, the DC-8 was used on the South American flights, and when Calder had finished with it, it looked like some kid in school had gone wild with his crayons on it. Not that it was awful—far from it. It was very intriguing and colorful. It caught everyone's attention and the ad campaign associated with it was quite successful. You have to say it was another stroke of genius by Harding Lawrence and Braniff. Braniff was not only an airline with a lot of vibrancy, vigor, and progressive "out there" ideas, but it was also successful at getting the audience to accept them. The rest of the airlines in the U.S. seemed very dour when you compared them to Braniff. The DC-8 was also showcased at one of the premier air show events in the world: the Paris Air Show. It was a head-turner there as well. What more could you ask? Touché, Lawrence.

There was another aircraft painted by Calder in 1976 for America's Bicentennial celebration in the colors of the U.S. flag. It was a 727-227, and the scheme was dubbed "Flying Colors of the United States." Braniff got extensive media coverage when Betty Ford, the First Lady of the United States, christened the aircraft at Washington Dulles airport on November 17, 1975. Sadly, the creator of the scheme, Alexander Calder, passed away almost a year later on November 11, 1976. At the time, he was working on a new scheme dubbed "Flying Colors of Mexico," which was never applied to any Braniff aircraft.

The Move to Dallas Fort Worth

In 1974, Braniff made a big move when it left its homely base at Dallas Love Field, which had been its base since 1941, and transferred to its new premises at Dallas Fort Worth International Airport. Its new base dwarfed the old one at 6,963 hectares, over five times the size of Dallas Love Field's 1,300 hectares. Dallas Fort Worth International Airport is America's second-largest airport, and another trivia fact: it is bigger than the island of Manhattan. Now, that is big. Braniff had a terminal of its own there: 2W or "The Braniff Terminal" as it was dubbed. The terminals were in semi-circles, and Braniff was the only airline to have a full semi-circle to itself. The other airlines were not so fortunate as each had only half a semi-circle as their terminal (the other airlines didn't have the traffic to warrant a full terminal). That included major operators like Eastern Airlines and American Airlines who had to share one, and Continental Airlines which shared another with Delta Airlines.

Ultra

By 1977, Braniff once again decided it needed to enhance its image further. Awash with cash after a fruitful year, with profits going through the roof, the carrier decided it needed a new livery. The long-serving designer, Emilio Pucci, was replaced by the American fashion designer Halston, who created the new Ultra look uniforms for the flight attendants. Moreover, together with Braniff's longtime industrial design firms, "Harper and George" of NYC, and the Detroit firm "Cars and Concepts," they designed the livery for the fleet of 727s, DC-8s, and 747s. The new look was unveiled on July 17, 1978, was quite subdued, and Elegant. Braniff called its latest campaign and look the "Elegance Ultra Campaign." It ushered in a new and sophisticated look that indicated Braniff had matured into one of the most magnificent airlines in the world.

It must be said that all was going well for Braniff in the mid to late 1970s. Its image, with its fleet of colorful airliners, was always going to turn heads and pull in customers. The company's finances were improving. Its new base was modern and futuristic, and a total of 16 new domestic destinations were offered to the public. Things were definitely looking up. However, on the horizon, plans were afoot which would forever change the face of civil aviation in the United States.

Deregulation

The calls for deregulating the airline industry had been going on for years. The protector of the major airlines and the allocation of routes, the CAB, was accused of bureaucratic complacency by many airlines, which had to wait for many years before they would be approved of their route requests or fare changes. This situation was exacerbated by the fact that the CAB often refused the applying airlines their requests, most of which had to wait many years before being rejected on trivial issues. Naturally, the major airlines with settled routes and year in and year out profits favored the system. Supplemental carriers such as Saturn Airways, World Airways, and even large scheduled carriers like Continental Airlines were victims of the system that the CAB had set up. In the 1970s, two major events—the 1973 oil crisis and stagflation—would show what was wrong with the system. When the oil crisis hit the airline world, the major carriers were guaranteed profits, but the flying public had to endure higher ticket prices. Many economists blasted the system and claimed that it promoted inefficiency, complacency, and higher costs. The idea of airline deregulation was first proposed during the Ford Administration and overlapped into Carter's Presidency. Thus, in 1977, Professor of Economics at Cornell University, Alfred Kahn, the antihero of the established carriers, was appointed by President Jimmy Carter to find the way forward This he did with, some would say, long-lasting negative effects that reverberate to this day. With support from leading economists, legislators, think tanks, and civil societies, the Airline Deregulation Act was passed, and was officially signed into law by President Carter on October 24, 1978. From that date on, many of the established carriers would constantly face stiff competition. Many would disappear completely, and others would either merge or allow themselves to be taken over by a rival entity. The decisions they would have to make to survive would need to be prescient and judicious. Anything less would spell disaster.

Braniff International's president, Harding Lawrence, was walking on a tightrope. He had to make the hard but correct decisions for the carrier after the advent of the Airline Deregulation Act. Many think he failed to see the dangers of deregulation and plunged, full of confidence, headfirst into the deep end, causing the ruin of Braniff International. Others might say he, nor any other airline President, could predict what the result of the ruling could mean for their airlines as there was no precedent. It was a game of Russian roulette.

In this period of deregulation, Lawrence was ready. He was the decision maker. He had to make the best decisions, and in his mind, he did. Here was the maestro—the flamboyant leader who oversaw many audacious advertising campaigns for Braniff, and the kaleidoscopic repainting of Braniff's fleet of airliners, splashed with vibrant color, dazzle, and pizzazz. He approved of the hostesses' "strips" and the stylish and colorful Braniff International customer lounges and ticket counters. He was always ahead of the curve when it came to creating a flamboyant image for his airline. His personality was overwhelming, he was the star of a movie he created and the people around him were his supporting actors. Everything about Lawrence pointed to a decision that was almost predictable.

The Post-Deregulation Expansion

Harding Lawrence did make the decision that many were expecting. It was to expand Braniff's network as far as he could, and to purchase many new airliners to assist in the expansion. It has to be said, Braniff International really had no choice but to expand, both domestically and internationally, so it could place itself in a position to be one of the few surviving carriers. This was because Southwest Airlines was offering stiff competition and eating away at Braniff's Texas feed. Had the expansion occurred in 1977, Lawrence might have been seen as a shrewd mastermind, and Braniff might have actually succeeded, where many other airlines had failed. Still it happened unknowingly, just as one of the worst recessions in US history was beginning.

Braniff was actually riding on a huge wave in the late 1970s and making good profits. For some, Lawrence's decisions were seen as too reckless. However, Braniff was not just another airline; it was one of the top ten airlines in the U.S. at the time. It had to take the bull by the horns and get through the problem of deregulation with an expansion that would get ahead of the opposition and improve its finances for the foreseeable future. The main changes in this expansion phase were the opening of 16 new routes to cities such as Los Angeles, Philadelphia, Salt Lake City, and Orlando, among others. It was in the international arena that he made the biggest splash. Los Angeles became a hub for Braniff for its flights to Asia, namely Hong Kong, Singapore, Seoul, and Guam. While, on the East Coast, Boston became Braniff's hub for its European route system, which comprised of Paris, London, Amsterdam, Frankfurt, and Brussels. Braniff had plans to extend flights to Tokyo, but they were not awarded, which negatively affected the Pacific Division and Bahrain, which were in fact awarded but not inaugurated.

A massive fleet expansion was put in force. New orders were made for new and leased aircraft. From the beginning of 1978 to 1981, the following types joined the fleet: twenty-five 727-200s, seven 747-100/200s (four of which were leased), and three 747SP-27s. The carrier ordered four 747SP-27, but in the end, it received only three. Braniff had prepared for the expansion, and for many years, since the early 1970s, had ordered ten new 727s per year, which replaced the older jets. This was still the case in 1979 when Braniff put in a single order of thirty 727s, but the plan was still to get ten 727s per year between 1979 and

1981, making it a total of thirty by 1981. However, the press erroneously reported that Braniff had bought all thirty 727s for delivery in one year

As for the 747s, they were ordered in 1978 and early 1979. The 747-100/200 fleet was mainly used on the European routes, and the 747SPs were placed on the Asian routes. The Asian sector was very profitable for Braniff, with almost full seating capacity, which was phenomenal. Initially, all load factors on Pacific and Atlantic routes were between 60 to 80 percent when they began in the summer and fall of 1979. This advantageous situation came to a sudden halt, when the Iran Hostage Crisis pushed fuel prices to unprecedented levels and caused them to exceed wage and salary costs as the highest airline expense, for the first time in aviation history. Then, with the global recession also appearing on the scene in the spring of 1980, load factors began dropping during the busy summer travel period of that year.

Concorde

Braniff had always had dreams of having a supersonic jetliner in their fleet. The carrier had reserved two spots in 1964 for US SST jet. It had also placed an order for 3 supersonic Concordes in 1966, which was unfortunately canceled in 1973. The desire, however, was still there to be the first carrier in America to go supersonic over American skies. They got a deal with British Airways and Air France to fly the awe-inspiring Concorde on their domestic Washington Dulles to Dallas Fort Worth route. With a great deal of interest, America tuned in on TV to the event of a lifetime when, on January 12, 1979, two Concordes, belonging to Air France and British Airways, flew in from Europe and landed at parallel runways at the same time at Dallas Fort-Worth International Airport.

Braniff had an interchange flight agreement with the two carriers, whereby the Braniff crews would fly the Concorde on the domestic sector from Washington Dulles to Dallas FW, and back. The European crews would then take over and fly the Concordes on the European sector. The Concordes kept their European liveries, and no Braniff "Concorde" was decked out in the Braniff's colors. However, The Concordes had to disguise their original country of registration code from "G" for British Airways, and "F" for Air France to the American code "N" temporarily while flying over the USA. This was because the FAA did not allow non-U.S. aircraft a U.S. certificate of airworthiness. The Concorde was limited to flying at Mach 0.95, but who could blame the crews for going supersonic at times, which they did. Finally, Braniff got its dream of being the first U.S. airline to fly a supersonic transport.

The feel-good factor with the Concorde did not last long. The carrier was making huge losses on the flights as the best they could achieve was only a 40% load factor, compared to the slower but more efficient 727-200s, which were basically full. The Concorde episode was not as successful as Braniff had hoped. The losses were nominal, but as fuel costs increased, so did the losses on the route. The service was stopped at the end of May 1980.

Financial Worries

Braniff International was financially in trouble by 1980. The higher costs of fuel, coupled with the CAB's and slow way of meandering through their application processes, hurt Braniff, and many other carriers. In Braniff's case, it lost $44 million in 1979 because it could not offset $60 million in fuel costs, by being allowed to raise fares. This was due to the CAB's slowness in deciding whether to allow the carrier to do so. The other reason was that airline discount coupons, were all used between Thanksgiving and

Christmas 1979. This caused Braniff's load factor to drop 5 points during the Fourth Quarter. From then on, Braniff would be fighting for its survival. Braniff's management tried to create a mega airline with a worldwide reach—it wanted to splash the world with the colors of Braniff and wanted to be the most successful airline in the new post-deregulation world. Alas, that was not to be. The new era of aviation had arrived. The way forward was by penny-pinching, offering a lower service, not listening to the passenger's grievances, and of course, paying staff lower salaries. This was, and is, the order of the day in the new dystopian era of deregulation.

Braniff was now looking for a way out of the malaise. The tragedy was that the carrier's finances were deteriorating at a very alarming rate. It thought about mergers and selling assets, but nothing came of those ideas. The debt was piling up, no airline could merge with Braniff in that condition, and the selling of assets would only alleviate the problem a little, but not enough to stave off the looming disaster.

Harding Lawrence was feeling the heat and he was requested by Braniff's board to retire, which he did in December 1981. John J. Casey then became the new President/CEO on January 8, 1981. He happened to be the older brother of Albert Casey, the American Airlines President. American Airlines was Braniff's fiercest rival as they duplicated Braniff's routes and offered more flights. On top of that, in 1979, they moved their base from New York to Dallas Fort Worth. The new CEO left Braniff after two months at the helm.

THE HOWARD PUTNAM ERA

The Final Stretch

The clock was ticking; time was not on Braniff's side. The management had to think and act fast. Howard Putnam was chosen as the new President in September 1981. He came in with excellent qualifications and experience. He was previously the President of Southwest Airlines, and a man with a level head on his shoulders. He got on the job of saving Braniff immediately. Most of the employees at the time saw him as the carrier's savior, and they were probably right. He tried as hard as anyone to turn things around.

Some of the measures that Putnam tried were tried and tested while he was at Southwest, and they did work. Braniff, however, was in a dire position financially, could his ideas work there? The first thing he did was to eliminate the first-class on all domestic flights coming from Southwest, that was a given. He did meet a lot of resistance from passengers about the loss of first-class, and in 1982, he was forced to introduce the new first-class named "Texas Class," though, with the looming bankruptcy on the horizon, it was futile. He slashed fares in half to attract the customers, and it did work. He made two other wise choices: he sold off all non-essential Braniff properties around the world and he stopped all European routes, except for one London flight per day. Controversially, he placed nearly a quarter of the staff members on furlough, which did not go down well at all, but there was no choice, it was either that or bankruptcy.

Putnam had to do something about American Airlines. As Braniff's arch-rival, he decided to take them on by starting a price war with its much bigger opponent. He introduced a giveaway sale to lure the customers. If they bought one round-trip ticket, they would get one for $1. The sale was a huge success and it annoyed American Airlines to the extent that they matched Braniff's sale, keeping the pressure on Braniff.

In March 1982, Putnam's other move was to sell all of Braniff's unprofitable routes to South America to its old rival, Pan American. The CAB was not convinced of the idea and said it needed to hold an oral evidentiary hearing. The decision took Braniff by surprise. The carrier did not have much time left. Braniff expected the decision by the CAB to let the sale go ahead—anything less would spell disaster.

In a last ditched effort, Putnam's team came up with another sale in April 1982. It was named "The great Braniff Escape Sale." They were confident that the sale would be a success, but sadly, that was not the case. The end was near.

The End

Braniff officially closed its doors at 12:01 am on May 13, 1982. The brave people, both at management level and the employees below them at Braniff, went through a lot in the two years leading up to the bankruptcy. It battled valiantly till the end, but it was not to be.

The end for Braniff, once America's eighth-largest airline, was caused by a combination of factors starting with deregulation, the ensuing expansion in the late 1970s, the bad economic environment in the early 1980s, and the fierce competition with major rivals. All these factors were enough to bring a proud airline like Braniff to its knees. One cannot blame Lawrence, for he did his best to continue the upward momentum of the carrier by expanding to new routes and purchasing new aircraft. Howard Putnam also did his best to keep Braniff flying. However, with his hands tied, while others, like the CAB, were making the decisions for Braniff, he was doomed to fail.

The founders of Braniff would have been looking on down from above that tragic May morning to see the end of their once great airline and would have wondered how it all could have been different. The answer to that question can never be known, but the Braniff name will always be remembered in history as Texas' greatest airline—a true Texan Giant.

SECTION 2

THE AIRCRAFT OF BRANIFF INTERNATIONAL AIRWAYS

1948 - 1982

CURTISS C-46 COMMANDO

Braniff International Airways operated two C-46 Freighters from 1955 to 1963. They were purchased from Civil Air Transport (C.A.T.) of Taiwan.

Braniff's one of two Curtiss C-46 Freighters, N1386N, was purchased in November 1955. It was used for freight services out of Dallas together with the other C-46, N1383N. *Digitized image by author.*

The second Braniff C-46 Freighter, N1383N, c/n 33641, was delivered in the same month as C-46 N1386N in 1955. Both C-46s served Braniff International until 1963, when they were sold to Air America. This aircraft crashed at Savannakhet, Laos, on November 25, 1968, whilst flying for Air America. *Photo courtesy of Pat Bukiri and the Midway Airport Historians.*

The C-46 Commandos were used as freighters for Braniff's cargo services. They were workhorses and provided good service for the carrier. *Digitized image by author.*

Profile view of the Braniff International C-46 Commando N1386N, c/n 22265. *Digitized image by author.*

CONVAIR 340/440

Braniff International was a major user of the Convair 340/440 prop liners. It operated a total of 43 aircraft, thirty-seven of which were the 340 model. Braniff also had a total of six CV-440 types in its fleet.

Braniff Convair CV-340-32 N3412, c/n 32, parked at Chicago Midway Airport in 1955. This aircraft was purchased new by Braniff and was delivered to the carrier on December 17, 1952. After 10 years of service, it was sold to Caribbean Atlantic Airlines in 1962. *Photo courtesy of Pat Bukiri and the Midway Airport Historians.*

Profile view of the Braniff International Convair CV-340-32 N3432, c/n 375. *Digitized image by author.*

Braniff Convair 340-32 N3407, c/n 20, at Chicago Midway Airport in 1958. It was delivered to Braniff on October 15, 1952. In November 1965, it was sold to Caribbean Atlantic Airways, and later converted to a freighter. It ended its flying days in 1995 and is now stored at Roswell Air Center Airport. *Photo courtesy of Pat Bukiri and the Midway Airport Historians.*

Convair CV-34-32, c/n 26, N3409 was delivered to Braniff International on November 29, 1952. It remained with the carrier until April 12, 1962, when it was sold to Caribbean Atlantic Airlines. *Photo courtesy of Pat Bukiri and the Midway Airport Historians.*

DOUGLAS DC-3

Braniff had a total of fourty-five DC-3s in its fleet over the airline's lifetime. Twenty-three of those were added to the Braniff International fleet after the takeover of Mid-Continent Airlines in 1952. The first DC-3 was delivered to Braniff on December 15,1939, and the last one was sold to Southern Airways on January 4, 1960. Twenty-one glorious years of service, with Braniff, for the legendary airliner.

DC-3 N61350, c/n 4535, was delivered to Braniff on August 15, 1952 and sold to Air Oasis on October 25, 1959. It is seen here, on a sunny day, at Chicago Midway Airport in 1955. *Photo courtesy of Pat Bukiri and the Midway Airport Historians.*

Profile view of the Braniff International Douglas DC-3 N61350, c/n 4535. *Digitized image by author.*

This is a photo of Braniff DC-3 N15201, c/n 11666, which was acquired after the takeover of Mid- Continent Airways in 1952. It was sold to Southern Airways on June 19, 1960. *Photo courtesy of Pat Bukiri and the Midway Airport Historians.*

Braniff International Douglas DC-3 N61350, c/n 4535. *Digitized image by author.*

LOCKHEED L-049 CONSTELLATION

Braniff International was not a big user of the Lockheed Constellation. It purchased two used "Connies" from the Venezuelan carrier Linea Aeropostal Venezolana – LAV, in August 1955. The L-049s were mainly used on domestic services to only four states: Texas, Oklahoma, Kansas, and Illinois. They were both eventually sold to Hubert M Covert in 1960 and 1961.

Braniff International Lockheed L-049 Constellation N2520B, c/n 2081, seen here parked at Chicago Midway Airport in 1957. The two Connies were nicknamed "Panic and "Fear" by the crews and maintenance guys, however, they were sound and reliable throughout their time with Braniff. *Photo courtesy of Pat Bukiri and the Midway Airport Historians.*

Profile view of the Braniff International Lockheed L-049 N2520B, c/n 2081. *Digitized image by author.*

Another photo of Braniff's L-049 N2520B. It is seen here at Chicago Midway Airport in the middle of winter in 1959. *Photo courtesy of Pat Bukiri and the Midway Airport Historians.*

Braniff International L-049 Constellation N2520B in flight. *Digitized image by author.*

DOUGLAS DC-4

Ten Douglas DC-4s were used by Braniff from1945, when it was still known as Braniff Airways, to 1954. All ten were ex USAAF aircraft that were converted to civil standard.

Braniff International Douglas DC-4 N65145, c/n 10360, was delivered to the carrier in November 1945. It was an ex USAAF transport and served Braniff until May 1954, when it was sold to Los Angeles Air Service. *Digitized image by author.*

Profile view of the Braniff International Douglas DC-4. *Digitized image by author.*

All ten DC-4s in service for Braniff International were ex USAAF surplus aircraft. *Digitized image by author.*

Braniff International DC-4 N65145. *Digitized image by author.*

DOUGLAS DC-6B

When the Douglas DC-6 came onto the scene in the late 1940s, it revolutionized air travel. It was one of America's greatest piston engine airliners, together with the graceful Lockheed Constellation series. Braniff initially purchased six DC-6, with the first one being delivered on August 23, 1947. It would eventually have eleven DC-6Bs in its fleet, 9 newly delivered from Douglas, and two used ones. The DC-6 fleet was eventually phased out by 1967.

N91311, c/n 45063, was the last DC-6 delivered to Braniff, which was in 1958. It was also the only 'B' model that Braniff had in its fleet. It is seen here at Chicago Midway Airport in 1958. It was sold to Lan Chile in October 1965. *Photo courtesy of Pat Bukiri and the Midway Airport Historians.*

Profile view of the Braniff International Douglas DC-6B N91311. *Digitized image by author.*

N90884, c/n 43108, was a DC-6 which was delivered to Braniff on October 28, 1947. After 19 years of service, it was sold to La Jolla Associates in February 1966. *Pho to courtesy of Pat Bukiri and the Midway Airport Historians.*

Braniff's DC-6 N90885, c/n 43109 is parked on its own in the late afternoon sun, at Chicago Midway Airport in 1960. *Photo courtesy of Pat Bukiri and the Midway Airport Historians.*

DOUGLAS DC-7C

The apogee in piston engine airline development was reached with the Douglas DC-7C. It was fast, it was sturdy, it was an airliner that could have flown for many years as a front-line transport. Unfortunately, it was rendered surplus by the introduction of the jetliners. Braniff's first DC-7C was delivered on September 13, 1956 and ended up having nine of them in the fleet. They were phased out by September 1970.

Braniff International DC-7C N5903, c/n 45071, is seen here on the tarmac at Chicago Midway Airport in 1959. It was with Braniff for 13 years before being sold to Club America Inc. on November 30, 1969. It was destroyed on October 9, 1986 in Dakar, Senegal, when it crashed after takeoff with an engine fire. *Photo courtesy of Pat Bukiri and the Midway Airport Historians.*

Profile view of the Braniff International Douglas DC-7C N9506, c/n 45074. *Digitized image by author.*

Braniff DC-7C N5906 was delivered on May 16, 1957. It was sold to Charlotte Aircraft Corporation in September 1970. *Photo courtesy of Pat Bukiri and the Midway Airport Historians.*

Braniff International Douglas DC-7C N5806. *Digitized image by author.*

LOCKHEED L-188 ELECTRA

Braniff International officially entered the turbo liner era on April 29, 1959, when the first of eleven Lockheed L-188 Electras joined the team. Ten years later, in April 1969, Braniff's Electra fleet was retired. They were all sold by the beginning of 1971.

This is a photo of Electra N9704C, c/n 1086, at Chicago Midway Airport circa 1959, not long after its delivery. Tragically, this aircraft crashed at Peru on December 24, 1971, whilst flying for Peruvian carrier LANSA. *Photo courtesy of Pat Bukiri and the Midway Airport Historians.*

Profile view of Braniff International L-188 N9704C, in Braniff's Pastel Blue colors. *Digitized image by author.*

Braniff International L-188 N9702C, c/n 1052, is seen here parked next to 727-27C N7277 at Chicago O'Hare, in June 1967.*Photo courtesy of the John F Ciesla Collection.*

Braniff International Lockheed L-188 N9704C, with its striking Pastel Blue livery, at Chicago O'Hare in October 1969.
Photo courtesy of the John F Ciesla Collection.

Braniff Airways Lockheed L-188 N9709C, c/n 1114, in the classic "El Dorado" scheme, at Chicago Midway Airport in 1961. *Photo courtesy of Pat Bukiri and the Midway Airport Historians.*

Profile of Braniff International L-188C N9704C in flight. *Digitized image by author.*

BAC 1-11

The BAC 1-11 was a British designed short-range twinjet, which made its first flight on August 20, 1963. On 20th October 1961, Charles Beard placed an order for six aircraft, with an option of six more. The first aircraft was delivered on March 11, 1965. The British jets replaced the CV-340/440 Convair props on regional routes.

Braniff International BAC-1-11 203AE N1546, c/n 20, in the early 1970s with the "Flying Colors" scheme. *Photo courtesy of the Eric Pajaud Collection.*

Profile view of Braniff International BAC 1-11 203AE N1550, c/n 44, in the Ochre scheme. *Digitized image by author.*

Braniff International BAC111-203-AE N1544, c/n 18, at Chicago O'hare, in June 1967. This aircraft served Braniff until July 1972, when it was sold to Allegheny Airlines. *Photo courtesy of the John F Ciesla Collection.*

Braniff BAC 1-11 N1554, c/n 71, taxiing at (Ronald Reagan) Washington National Airport in May 1966. *Photo courtesy of the John F Ciesla Collection.*

BOEING 707/720

The King of the classic jetliners, the Boeing 707, was chosen by Braniff quite early. In 1955, Charles Beard put in an order for 5 Boeing 707-227s. Braniff was the only operator to choose this model. It was a wise choice as Braniff became one of the few airlines in the world to have turbojet airliners in its fleet at the end of 1959. Later, it also chose the 720-027, the 707-138B, and the 707-327C. This was the beginning of a love affair by Braniff and Boeing, which supplied most of the jetliners for the BI fleet over the years.

Braniff's classic 707-227 N7073, c/n 17693, at Idlewild in June 1960. When the 707-227 was delivered, it was the fastest jet airliner in the world, making Braniff the "fastest airline in the world", at the time! *Photo courtesy of the John F Ciesla Collection.*

Braniff 707-227 at Idlewild Airport in September 1960. The 707-227 was powered by four Pratt & Whitney JT4A-3 turbojets, compared to the 707-120 which was using the less powerful Pratt & Whitney JT3C-6 turbojets. *Photo courtesy of the John F Ciesla Collection.*

Braniff International 720-027 N7079, c/n 18423, Orange "Jellybean" taking off. *Photo courtesy of the Trevor Ogle Collection.*

Braniff International 707-327C N7097, c/n 19106, in the Light Green livery at Honolulu Airport in 1971.
By RuthAS
https://commons.wikimedia.org/w/index.php?search=RuthAS+braniff&title=Special%3ASearch&go=Go&ns0=1&ns6=1&ns12=1&ns14=1&ns100=1ns106=1#/media/File:Boeing_707-327C_N7097_BN_HNL_21.04.71_edited-3.jpg
https://creativecommons.org/licenses/by/3.0/#. Cropped from original.

Profile of Braniff International 707-327C N7100, c/n 19440, in the Green "Jellybean" livery. *Digitized image by author.*

DOUGLAS DC-8

Braniff was an operator of Douglas piston engine airliners, but in the arrival of the jetliners, Braniff swiftly switched allegiance to Boeing. After the Panagra merger, it operated the DC-8-31, DC-8-51 & 55F, and the DC-8-62 series, although they were mainly used airliners. The only brand-new DC-8s they purchased from the Douglas Corporation were two DC-8-62s, in 1969.

Braniff International DC-8-62 N1804, c/n 45896, ex Panagra, in the "Jellybean" scheme, at JFK Airport New York in February 1969. *Photo courtesy of the John F Ciesla Collection.*

Profile view of the Braniff International DC-8-62 N1808E, c/n 46105, in the "Flying Colors" livery. *Digitized image by author.*

Braniff International DC-8-62 N1808E, c/n 46105, with the "Flying Colors" scheme, landing at Miami International. *Photo courtesy of George Mihalek.*

Braniff International DC-8-62 N1806, c/n 45911, in the "Ultra" livery taking off. *Photo courtesy of the Trevor Ogle Collection.*

Braniff DC-8-62 N802BN, c/n 45909, at Los Angeles International in November 1980. *Photo courtesy of George Mihalek.*

Braniff DC-8-62 N810BN, c/n 45905, landing at Miami International in January 1980. *Photo courtesy of George Mihalek.*

BOEING 727

If there is one airliner type that would epitomize Braniff International, America's sleekest airline, it would be the Boeing 727. It has to be Boeing's sleekest airliner ever. It was a classy airliner, which, at one stage in the early 1980s, was also Boeing's biggest selling airliner. Braniff operated a total of 131 Boeing 727s, which is a huge number. For anyone who thinks of Braniff nowadays, they also think of the 727.

Braniff International 727-027C N7280, c/n 19119, at Chicago O'Hare in July 1967. *Photo courtesy of the John F Ciesla Collection.*

Profile of Braniff International 727-027C N7289, c/n 19499, in the yellow "Jellybean" color. *Digitized image by author.*

Braniff International 727-027C N7289 Yellow "Jellybean". *Digitized image by author.*

Braniff International 727-027C N7281, c/n 19120, Green "Jellybean" at Chicago O'Hare in September 1967. *Photo courtesy of the John F Ciesla Collection.*

Braniff International 727-027 N7289, c/n 19499, landing at Chicago O'Hare in July 1977. *Photo courtesy of George Mihalek.*

Braniff International 727-191 N298BN, c/n 19392, at Cleveland Hopkins International in October 1975. *Photo courtesy of George Mihalek.*

Braniff International 727-227 N439BN, c/n 2118, on finals at Cleveland Hopkins International in April 1980. *Photo courtesy of George Mihalek.*

Braniff International 727-227 N403BN, c/n 20394, at the gate at Chicago O'Hare in November 1976. *Photo courtesy of George Mihalek.*

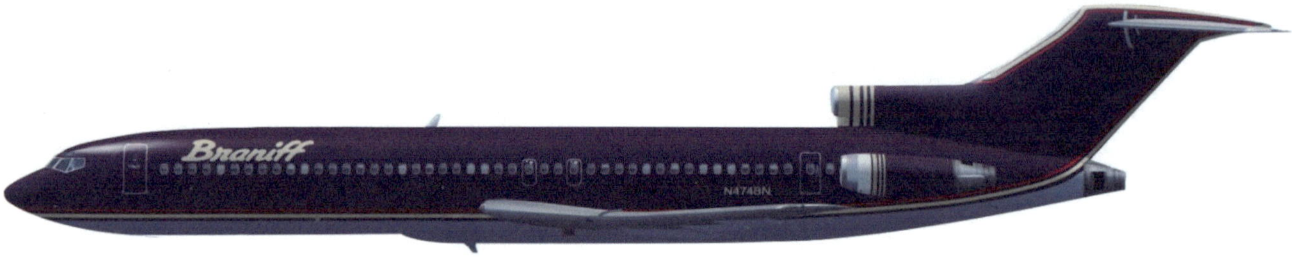

Profile of Braniff International 727-227 N474BN, c/n 21997, in the "Ultra" colors. *Digitized image by author.*

Braniff International 727-227 N474BN, c/n 21997 in the Chocolate Brown "Ultra" livery.The airliner was delivered to Braniff on January 29, 1980. *Photo courtesy of George Mihalek.*

Braniff 727-227 N471BN, c/n 21669, on finals at Los Angeles International in the Perseus Green "Ultra" livery in October 1980. *Photo courtesy of George Mihalek.*

Braniff 727-227 N457BN, c/n 21463, taxiing at Tampa International Airport. *Photo courtesy of* the *Trevor Ogle Collection.*

BOEING 747/747SP

Braniff International received its first Boeing 747, registered N601BN, on January 1, 1971. It was painted all orange and it was a sight to see at the airport. The "Big Orange," or "The Great Pumpkin," as it was dubbed, was extremely popular. It was mainly used on flights to Hawaii, in the early to mid-1970s, and it was very profitable for Braniff. In the late 1970s, there was a huge boost in purchases of the 747 jetliners, a total of ten joined the fleet between 1978 and 1980. Five of those were bought by the carrier, and five were leased.

Braniff "Big Orange" 747-227B N602BN, c/n 21682, at Amsterdam, The Netherlands. *Photo courtesy of Maarten Visser.*

Profile of Braniff 's first Jumbo, 747-127 SCD N601BN, c/n 20207, "Big Orange". *Digitized image by author.*

Profile of Braniff International 747SP-27 N606BN, c/n 21992. *Digitized image by author.*

Braniff International 747SP-27 at Los Angeles International Airport, 1981. By clipperarctic. https://commons.wikimedia.org/wiki/File:Braniff_747SP_(7491681708).jpg. "Braniff 747SP (7491681708)", https://creativecommons.org/licenses/by-sa/2.0/ Enhanced saturation.

THE CALDER AIRCRAFT

The famous artist, Alexander Calder, was commissioned by Braniff International to create some to paint a DC-8-51 in the "Flying Colors of South America" scheme. In a first for any airline, Braniff created an image of surreal excellence. The paint on the DC-8 was also displayed at the 1973 Paris Air Show, and it was an attention grabber there. It was colorful, it was vibrant... it was Braniff! Alexander Calder also painted a 727-200 in the "Flying Colors of the United States" in 1976, to celebrate America's Bicentennial.Braniff got extensive media coverage when Betty Ford, the First Lady of the United States, christened the aircraft at Washington Dulles airport on November 17, 1975. Sadly, Alexander Calder passed away on November 11, 1976. At the time, he was working on a new scheme dubbed "Flying Colors of Mexico," which was never finished.

Braniff DC-8-62 N1805, c/n 45899, in the Calder "Flying Colors of South America" scheme at Cleveland Hopkins International in April 1977. *Photo courtesy of George Mihalek.*

Braniff DC-8-62 N1805, 'Calder' "Flying Colors of South America" on short final at Miami International in March 1978. *Photo courtesy of George Mihalek.*

Braniff International 727-291 N408BN, c/n 19993, in the 'Calder' "Flying Colors of the United States" scheme at Cleveland Hopkins International. *Photo courtesy of George Mihalek.*

BRANIFF TIMETABLE IMAGES

THROUGH THE AGES

These images are from Björn Larsson, and his informative website: https://www.timetableimages.com/

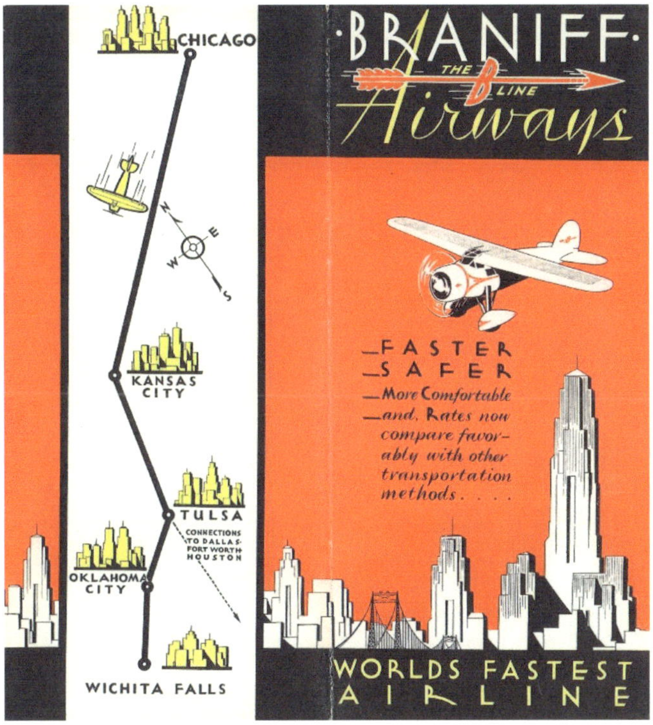

1931

1939

1945

1948 - June

1948 - June

1948 - February

1948

Fly BRANIFF
Through the Houston Air Gateway Braniff flies three times a week to Havana, Cuba, Balboa-Panama City, Guayaquil, Ecuador and Lima, Peru . . . twice a week to La Paz, Bolivia . . . twice a week, non-stop from Lima, Peru to Rio de Janeiro, Brazil. Braniff flights within the United States serve 32 mid-west cities and through well-timed connections link them with every major city in the United States and with Braniff's expanding South American system.
SCHEDULES AND FARES APRIL 1, 1949
BRANIFF International AIRWAYS
BRANIFF
Now Fly BRANIFF to RIO
ASK ABOUT BRANIFF ALL EXPENSE AIR-TOURS

1949

to
south america
Where international trade's at a new high and vacations were never more fabulous or more fun! Why not combine business with pleasure and fly south with us to Havana, Panama, Lima, Sao Paulo, Rio and Buenos Aires? Go elegantly on one of our deluxe DC-6 sleeper flights—or economically aboard a fine 4-engine tourist liner. Costs less than you think, too. For instance, you can go round trip round South America for as low as $912.00 from the Miami Gateway.
it's BRANIFF
INTERNATIONAL AIRWAYS
For information and reservations, call your travel agent or any of Braniff's 139 offices throughout the hemisphere.
Schedules Effective July 1, 1953
All schedules shown in local time
it's BRANIFF
INTERNATIONAL AIRWAYS
T. E. BRANIFF, PRESIDENT AND CHAIRMAN OF THE BOARD
BRANIFF INTERNATIONAL AIRWAYS
Observing Twenty-five Years of Public Service

BRANIFF
1928
1953
This is our silver anniversary year. Back in 1928, three Braniff pioneers, with a four-passenger Stinson started flying a 116-mile route, from Oklahoma City to Tulsa.
Observing Twenty-five Years of Public Service
That was the beginning. Since then, we've grown and helped others grow. We introduced I.L.S., JATO and "instant reservations". Then, we inaugurated the world's longest overland flight—non-stop from Lima to Rio. Recently, we added still another important first when we launched the new Super Convair 340.
We feel we've come a long way in improving our service and expanding our routes through nine countries in this hemisphere. But it's no time for resting on laurels. Instead, we're renewing our pledge to deliver the finest in flying the Americas!
it's BRANIFF
INTERNATIONAL AIRWAYS
Love Field • Dallas, Texas
AIR MAIL • AIR PARCEL POST • AIR FREIGHT • AIR EXPRESS

1953

SEPTEMBER OCTOBER NOVEMBER
FAMILY-FARE PLAN
Braniff gives budgets a break with this popular family flying plan. The head of the family, either husband or wife, pays full fare, and the spouse and/or children over 12 and under 22 years of age pay only two-thirds fare when travel starts at any time Tuesday, Wednesday, or Thursday, (one accompanied child under two flies free, those under 12 pay only half-fare anytime.)
You can leave on a Braniff flight as late as one minute before midnight Thursday and continue traveling Friday on Braniff and most other carriers at family-fare plan rates.
The whole family leaves on the same flight, but it's not necessary for all to return on one flight, so long as return days are Tuesday, Wednesday and Thursday. If Dad is pressed for time, the others can come back later.
BRANIFF International AIRWAYS
PRINTED IN U.S.A.
Effective October 1, 1959
BRANIFF International AIRWAYS
Largest, Oldest Airline Serving the Heart of Both Americas
FLY BRANIFF'S
Jet Power
ELECTRA
PACIFIC Standard time
MOUNTAIN Standard time
CENTRAL Standard time
EASTERN Standard time
TABLE OF CONTENTS
TICKET OFFICES
DOMESTIC SCHEDULES
INTERNATIONAL SCHEDULES
PASSENGER INFORMATION
THRU ONE-PLANE SERVICE TO SOUTH AMERICA
BRANIFF INTERNATIONAL AIRWAYS SERVICE IN SOUTH AMERICA AND TO THE UNITED STATES
BRANIFF AIRWAYS CERTIFICATED ROUTES
CONNECTING AIRLINES
INTERCHANGE SERVICE
BRANIFF International AIRWAYS
serving more major cities in the U.S.A. and South America . . . than any other airline!

1959

September October November
BUSINESSMEN'S SPECIAL
EL DORADO SUPER JETS
NON-STOP
BETWEEN
CHICAGO
HOUSTON
Now you can truly relax on Braniff's new "Businessmen's Special." Read the latest men's, business and general publications as you enjoy a cocktail, mixed drink or a cold, refreshing beer. Lean back and light up your favorite pipe or a mellow cigar. Complimentary liquor, beer and cigars are featured in the Deluxe First Class compartment, and are available at nominal cost in the tourist section.
BRANIFF International AIRWAYS
PRINTED IN U.S.A.
Effective September 3, 1963
BRANIFF International AIRWAYS
SYSTEM TIMETABLE
NOW! 5
SUPER JETS
NEW YORK • DALLAS
PACIFIC Standard time
MOUNTAIN Standard time
CENTRAL Standard time
EASTERN Standard time
TABLE OF CONTENTS
THRU ONE-PLANE SERVICE TO SOUTH AMERICA VIA BRANIFF/EASTERN
BRANIFF INTERNATIONAL AIRWAYS SERVICE IN SOUTH AMERICA AND TO THE UNITED STATES
BRANIFF AIRWAYS CERTIFICATED ROUTES
CONNECTING AIRLINES
INTERCHANGE SERVICE
BRANIFF International AIRWAYS
Best in the air for personal care

1963

Fly Braniff to New York and the World's Fair
Ask about new low cost all-inclusive tours. Prices start as low as $19.95 plus air fare.
Six jet flights from Dallas . . . three non-stop. Excellent connections at Dallas from all the Southwest.
For information or reservations contact your travel agent or call Braniff.
Braniff International
United States Mexico South America
PRINTED IN U.S.A.
Braniff International
United States Mexico South America
Braniff International
United States Mexico South America
Braniff International
United States Mexico South America
Braniff International
United States Mexico South America
Braniff International
United States Mexico South America
System Timetable Effective June 1, 1965
Braniff International Routes and Principal Connecting Services
United States Mexico South America
To HAWAII & the ORIENT
THRU PLANE SERVICE London and Frankfurt With Pan American Airways
THRU PLANE SERVICE Seattle and Portland With United Air Lines
To EUROPE
THRU PLANE SERVICE New York-South America With Eastern Air Lines
to leading U. S. cities or South America
Specialists in international jet service
SEATTLE
PORTLAND
MINNEAPOLIS ST. PAUL
DENVER
OMAHA
CHICAGO
KANSAS CITY
NEW YORK NEWARK
WASHINGTON
DALLAS
HOUSTON
SAN ANTONIO
MIAMI
MEXICO CITY
ACAPULCO
LOS ANGELES
EL PASO
BOGOTA
QUITO
GUAYAQUIL
LIMA
LA PAZ
RIO DE JANEIRO
SAO PAULO
ASUNCION
BUENOS AIRES
TABLE OF CONTENTS
Page
TICKET OFFICES . . . 2
PASSENGER INFORMATION . . . 3
DOMESTIC SCHEDULES
Quick Reference Schedules . . . 4-5
New York - Washington - Dallas - South and West Texas - Mexico City . . . 7 8 9
Minneapolis - St. Paul - Chicago - Denver - Texas - Mexico City - Louisiana and Intermediate Points
Southbound . . . 10-11-12-13
Northbound . . . 14-15-16-17
Interchange Schedules . . . 18
Domestic Fares . . . 19
INTERNATIONAL SCHEDULES
Southbound . . . 20
Northbound . . . 21
International Fares . . . 22
BRANIFF AIRWAYS CERTIFICATED ROUTES
THRU PLANE INTERCHANGE SERVICE
New York-South America: With Eastern Air Lines
London and Frankfurt — Southwest Cities: With Pan American Airways
Seattle and Portland — Southwest Cities: With United Air Lines
Denver-St. Louis: With Continental Airlines
Oklahoma and Arkansas Cities — Birmingham and Atlanta: With Eastern Air Lines
CONNECTING AIRLINES
NOTE: This map does not show the full network of BRANIFF's nonstop routes.

1965

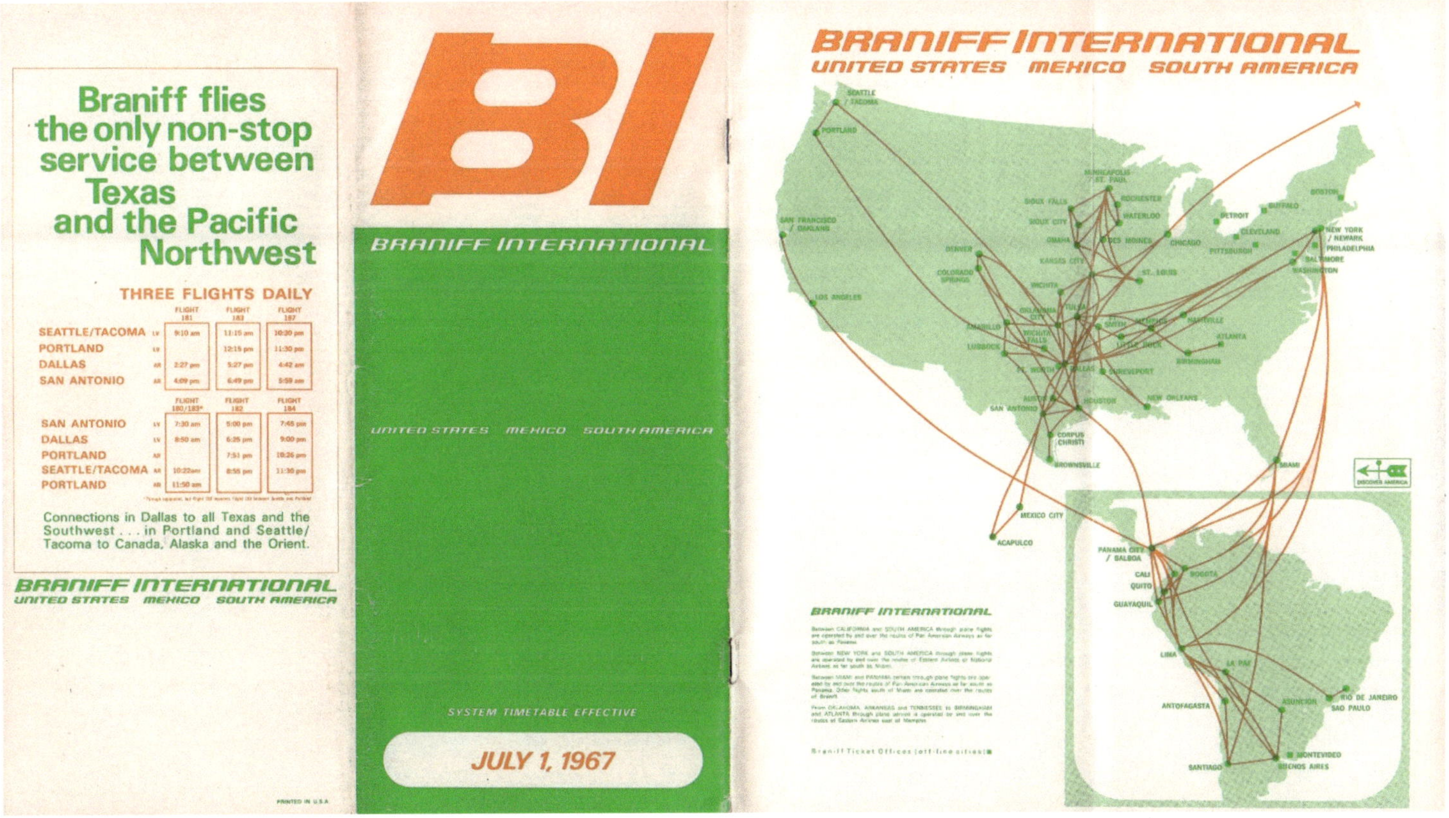
Braniff flies the only non-stop service between Texas and the Pacific Northwest
THREE FLIGHTS DAILY
FLIGHT 181 / FLIGHT 183 / FLIGHT 187
SEATTLE/TACOMA LV 9:10 am 11:15 am 10:30 pm
PORTLAND LV 12:15 pm 11:30 pm
DALLAS AR 2:27 pm 5:27 pm 4:42 am
SAN ANTONIO AR 4:09 pm 6:49 pm 5:59 am
FLIGHT 180/183* / FLIGHT 182 / FLIGHT 184
SAN ANTONIO LV 7:30 am 5:00 pm 7:45 pm
DALLAS LV 8:50 am 6:25 pm 9:00 pm
PORTLAND AR 7:51 pm 10:26 pm
SEATTLE/TACOMA AR 10:22am 8:55 pm 11:30 pm
PORTLAND AR 11:50 am
Connections in Dallas to all Texas and the Southwest . . . in Portland and Seattle/Tacoma to Canada, Alaska and the Orient.
BRANIFF INTERNATIONAL
UNITED STATES MEXICO SOUTH AMERICA
PRINTED IN U.S.A.
BI
BRANIFF INTERNATIONAL
UNITED STATES MEXICO SOUTH AMERICA
SYSTEM TIMETABLE EFFECTIVE
JULY 1, 1967
BRANIFF INTERNATIONAL
UNITED STATES MEXICO SOUTH AMERICA
SEATTLE / TACOMA
PORTLAND
SAN FRANCISCO / OAKLAND
LOS ANGELES
MINNEAPOLIS / ST. PAUL
DENVER
CHICAGO
NEW YORK / NEWARK
WASHINGTON
HOUSTON
SAN ANTONIO
MIAMI
MEXICO CITY
ACAPULCO
PANAMA CITY / BALBOA
BOGOTA
QUITO
GUAYAQUIL
LIMA
LA PAZ
ANTOFAGASTA
SANTIAGO
BUENOS AIRES
ASUNCION
RIO DE JANEIRO
SAO PAULO
MONTEVIDEO
DISCOVER AMERICA
BRANIFF INTERNATIONAL

1967

When you land, take off again with National Car Rental
IN A HURRY? Renting a car at National is fast, simple, pleasant.
We'll have you in a fresh new Chevrolet or other fine GM car in seconds.
If you don't have a National Credit Card, we'll honor any other recognized credit card until you get one.
Let us call ahead for a reservation. We'll have a computerized confirmation for you at any of over 1350 locations . . . in 5 seconds!
And when you turn your car in, we say "thanks" with free S&H Green Stamps for every mile you've driven.
Wherever you are, rent a National car.
NATIONAL CAR RENTAL
where the customer is always No. 1
PRINTED IN U.S.A.
BI
BRANIFF INTERNATIONAL
UNITED STATES MEXICO SOUTH AMERICA
SYSTEM TIMETABLE EFFECTIVE
JULY 1, 1968

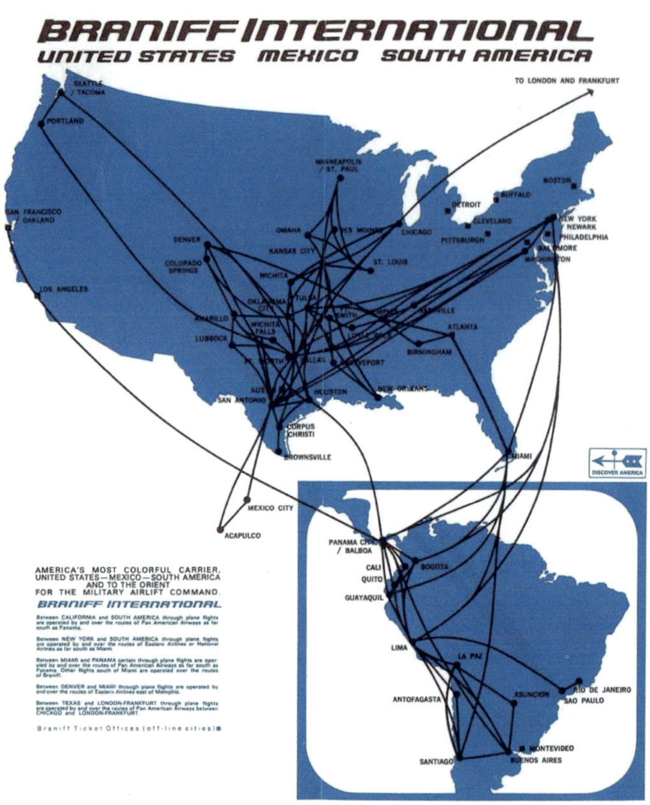
BRANIFF INTERNATIONAL
UNITED STATES MEXICO SOUTH AMERICA
TO LONDON AND FRANKFURT
SEATTLE / TACOMA
PORTLAND
SAN FRANCISCO OAKLAND
LOS ANGELES
DENVER
COLORADO SPRINGS
MINNEAPOLIS / ST. PAUL
OMAHA
KANSAS CITY
WICHITA
ST. LOUIS
CHICAGO
DETROIT
BUFFALO
CLEVELAND
PITTSBURGH
BOSTON
NEW YORK / NEWARK
PHILADELPHIA
WASHINGTON
LUBBOCK
ATLANTA
BIRMINGHAM
HOUSTON
NEW ORLEANS
SAN ANTONIO
CORPUS CHRISTI
BROWNSVILLE
MIAMI
MEXICO CITY
ACAPULCO
DISCOVER AMERICA
PANAMA CITY / BALBOA
CALI
QUITO
GUAYAQUIL
BOGOTA
LIMA
LA PAZ
ANTOFAGASTA
ASUNCION
RIO DE JANEIRO
SAO PAULO
MONTEVIDEO
BUENOS AIRES
SANTIAGO
AMERICA'S MOST COLORFUL CARRIER, UNITED STATES—MEXICO—SOUTH AMERICA AND TO THE ORIENT FOR THE MILITARY AIRLIFT COMMAND.
BRANIFF INTERNATIONAL
Between CALIFORNIA and SOUTH AMERICA through plane flights are operated by and over the routes of Pan American Airways as far south as Panama.
Between NEW YORK and SOUTH AMERICA through plane flights are operated by and over the routes of Eastern Airlines or National Airlines as far south as Miami.
Between MIAMI and PANAMA certain through plane flights are operated by and over the routes of Pan American Airways as far south as Panama. Other flights south of Miami are operated over the routes of Braniff.
Between DENVER and MIAMI through plane flights are operated by and over the routes of Eastern Airlines east of Memphis.
Between TEXAS and LONDON-FRANKFURT through plane flights are operated by and over the routes of Pan American Airways between CHICAGO and LONDON-FRANKFURT.
Braniff Ticket Offices (off-line cities)

1968

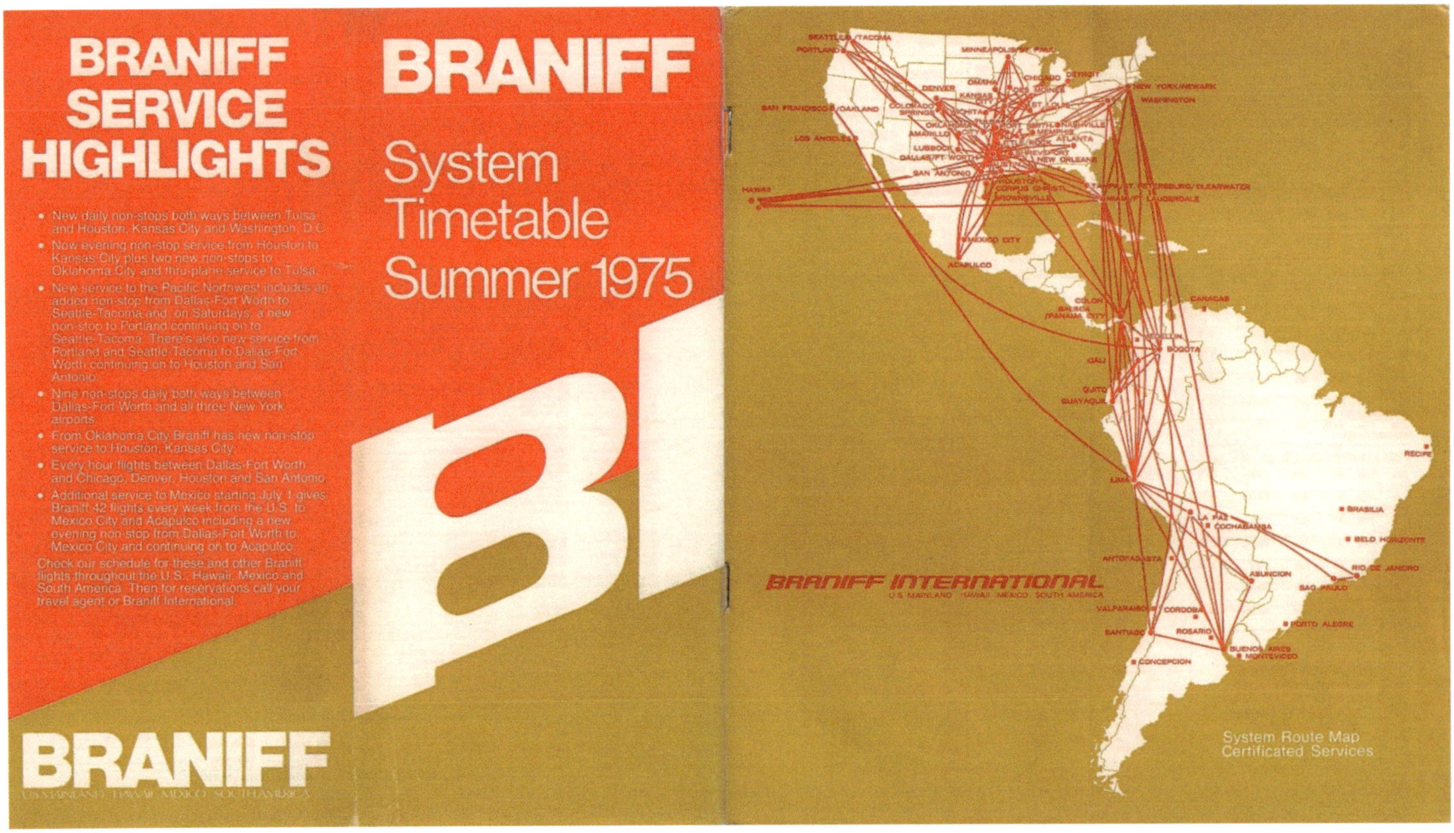
BRANIFF SERVICE HIGHLIGHTS
• New daily non-stops both ways between Tulsa and Houston, Kansas City and Washington, D.C.
• Now evening non-stop service from Houston to Kansas City plus two new non-stops to Oklahoma City and thru-plane service to Tulsa.
• New service to the Pacific Northwest includes an added non-stop from Dallas-Fort Worth to Seattle-Tacoma and, on Saturdays, a new non-stop to Portland continuing on to Seattle-Tacoma. There's also new service from Portland and Seattle-Tacoma to Dallas-Fort Worth continuing on to Houston and San Antonio.
• Nine non-stops daily both ways between Dallas-Fort Worth and all three New York airports.
• From Oklahoma City Braniff has new non-stop service to Houston, Kansas City.
• Every hour flights between Dallas-Fort Worth and Chicago, Denver, Houston and San Antonio.
• Additional service to Mexico starting July 1 gives Braniff 42 flights every week from the U.S. to Mexico City and Acapulco including a new evening non-stop from Dallas-Fort Worth to Mexico City and continuing on to Acapulco.
Check our schedule for these and other Braniff flights throughout the U.S., Hawaii, Mexico and South America. Then for reservations call your travel agent or Braniff International.
BRANIFF
BRANIFF
System Timetable Summer 1975
BI
BRANIFF INTERNATIONAL
System Route Map Certificated Services

1975

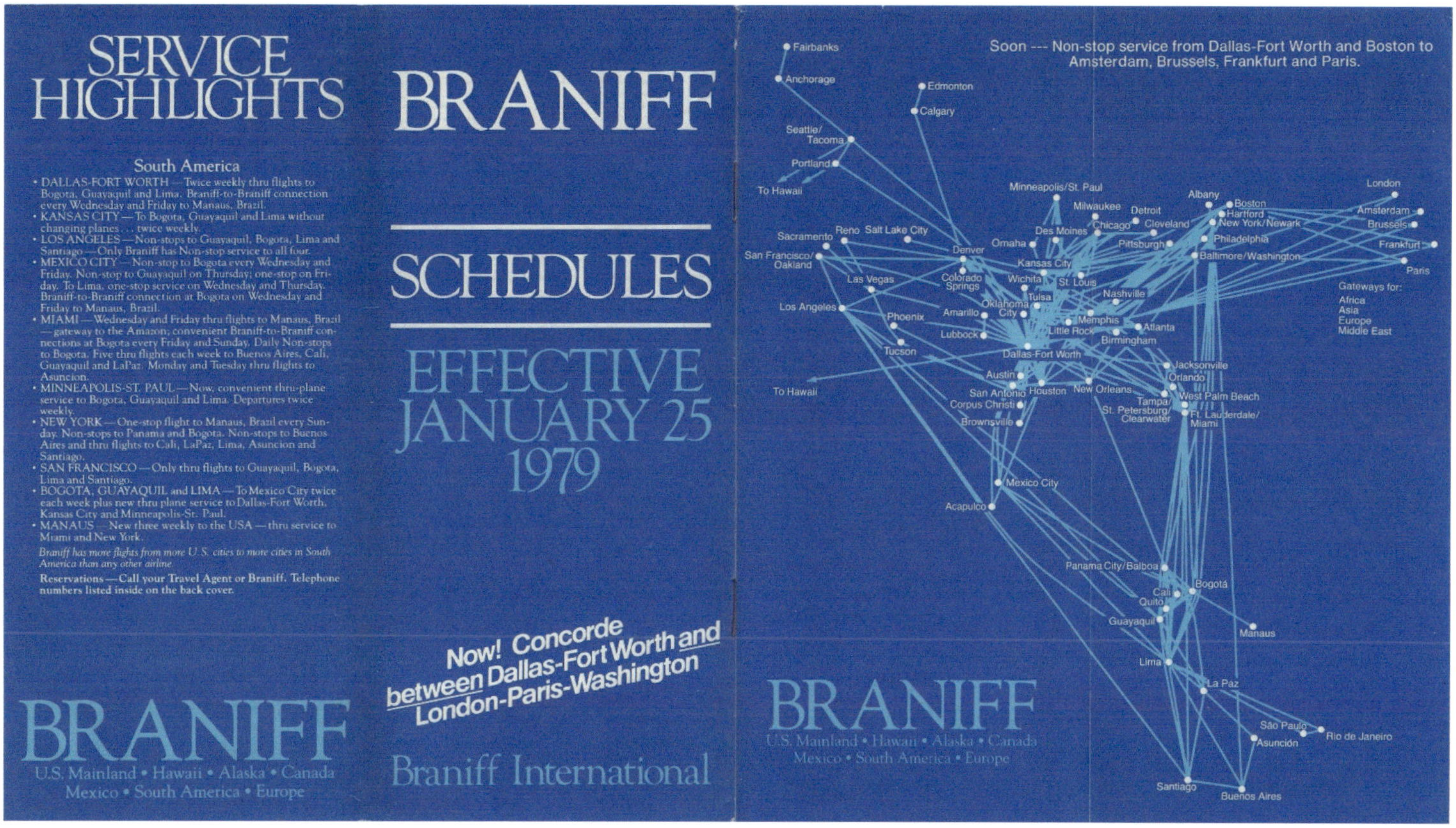

1979 – January

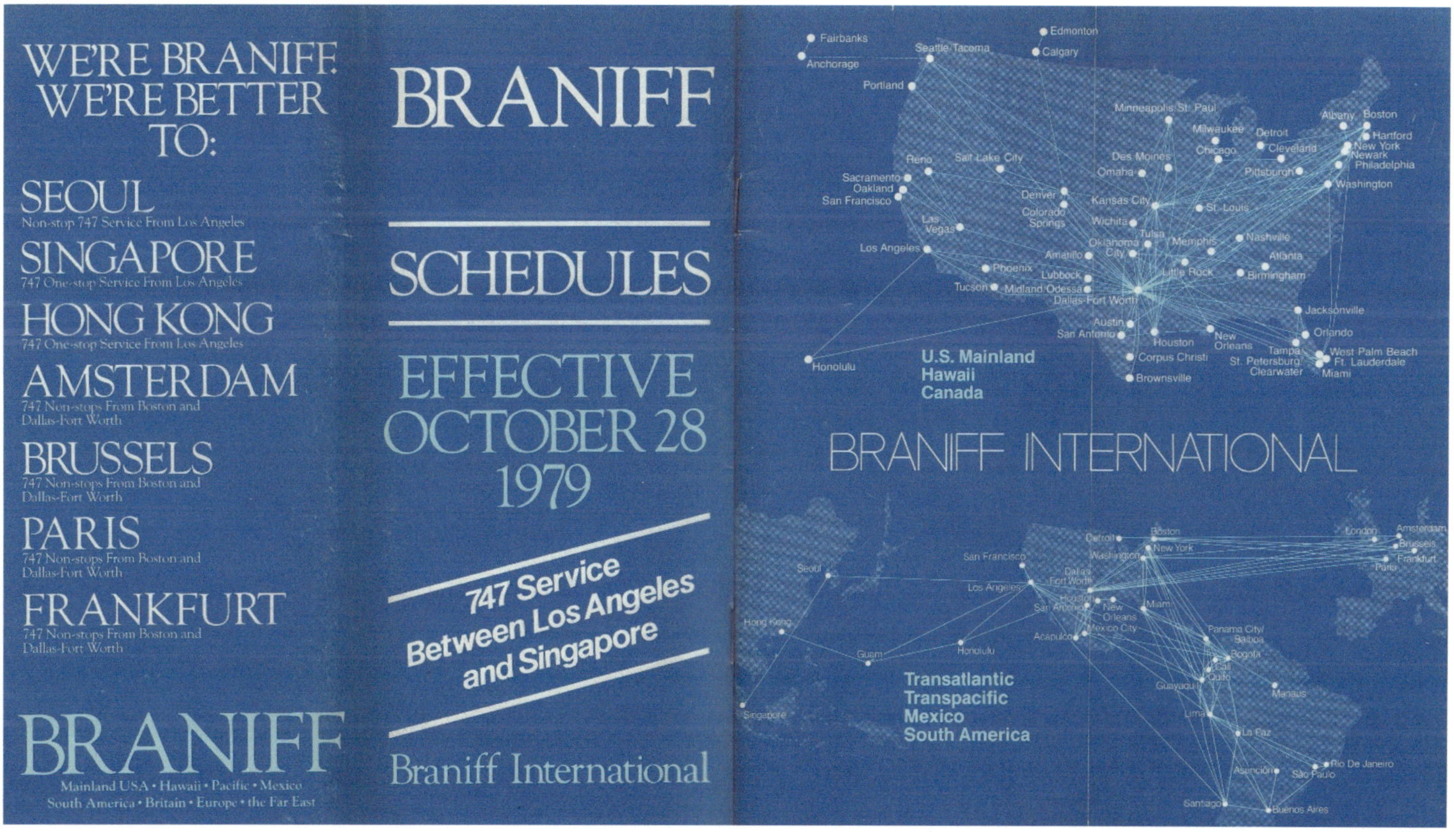

1979 – October

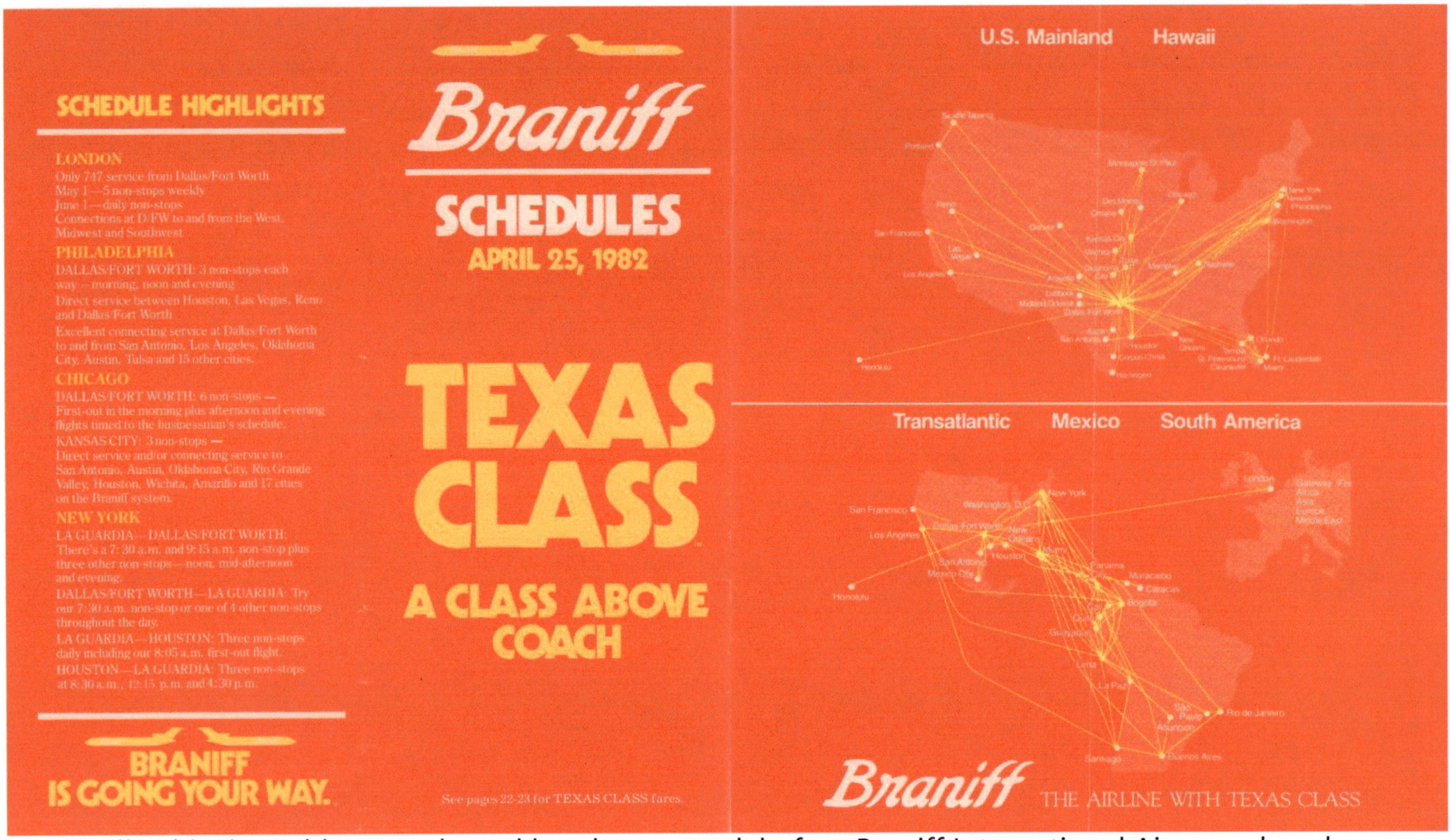

Sadly, this timetable was released barely one week before Braniff International Airways closed its doors on May 2, 1982.

BRANIFF INTERNATIONAL AIRWAYS FULL HISTORIC FLEET 1948 – 1982

BRANIFF INTERNATIONAL AIRWAYS HISTORIC FLEET - REFERENCE 1948-1982

Abbreviations: WFU: Wihdrawn from use ST: Sold To BU: Broken up LSD: Leased MRG: Merged BT: Bought OB: On Board FT: Fatalities RG: Registered STD: Stored DBR: Damaged Beyond Repair SCR: Scrapped RTD: Returned CV: Converted

C/N	REG	Date Delivered	Type	REMARKS
	BAC 1-11			
15	N1541	8/10/65	111-203AE	Braniff BT new. ST Omni 10/16/72.
16	N1542	4/20/65	111-203AE	Braniff BT new. ST Allegheny Airlines 8/18/1972. BU KMIA 8/87
17	N1543	3/11/65	111-203AE	Braniff BT new. ST McCulloch International Airlines 3/77. WFU STD Lagos, Nigeria 2/98. BU 2002
18	N1544	4/6/65	111-203AE	Braniff BT new. ST Allegheny Airlines 7/72. WFU STD Hondo TX. BU 1997
19	N1545	5/12/65	111-203AE	Braniff BT new. ST Allegheny Airlines 7/72. WFU STD Hondo TX. CNC 11/12/97
20	N1546	6/2/65	111-203AE	Braniff BT new. ST Allegheny Airlines 9/7/72. BU KMCO 9/91
41	N1547	7/20/65	111-203AE	Braniff BT new. ST Allegheny Airlines 4/72. WO Aircraft skidded off runway on landing, Lagos-Calabar Nigeria, 7/29/97. OB:55 FT:1
42	N1548	8/18/65	111-203AE	Braniff BT new. ST Allegheny Airlines 5/4/72. BU KHDO. CNC 10/1/97
43	N1549	9/24/65	111-203AE	Braniff BT new. ST Allegheny Airlines 4/17/72. WFU STD EGMC. BU Southend EGMC 7/96
44	N1550	10/4/65	111-203AE	Braniff BT new. WO Pilot came in too fast and high with a nose down attitude. Plane hit runway hard, and overran it, striking a draining ditch 7/9/78. No Fatalities.
45	N1551	11/8/1965	111-203AE	Braniff BT new. ST Mohawk AL 9/21/70. WFU STD KSAT
46	N1552	11/24/65	111-203AE	Braniff BT new. ST Mohawk AL 9/15/70. WFU CNC 03/28/96. BU Hondo TX 1997
70	N1553	12/8/65	111-203AE	Braniff BT new. WO Aircraft suffered in-flight structural failure caused by extreme turbulence during a thunderstorm. Crashed near Falls City NE 8/6/66. OB:42 FT:42
71	N1554	12/22/65	111-203AE	Braniff BT new. ST Mohawk AL 9/70. WFU Stored at Hondo Texas Dec. 1982
	Boeing 707			
17692	N7072	12/3/59	707-227	Braniff BT new. ST BWIA 2/1/71. BU KMIA 1987
17693	N7073	1/15/60	707-227	Braniff BT new. ST BWIA 5/11/71. BU TLV 9/77
17694	N7074	1/21/60	707-227	Braniff BT new. ST BWIA 3/27/71. BU KMWH 9/81, CNC 12/4/84
17695	N7075	2/10/60	707-227	Braniff BT new. ST BWIA 4/10/71. STD 3/75, BU for parts 1977

18068	N105BN	9/28/69	707-138B	Ex QANTAS ST Boeing Company 11/75. BU 1991
18069	N106BN	7/5/69	707-138B	Ex QANTAS ST Montana Austria 7/26/77.
18739	N107BN	6/19/69	707-138B	Ex QANTAS LSD to Merpati Indonesia 7/15/76. WFU10/88 at SNN. BU 1993, at CNC 12/22/14
18740	N108BN	6/24/69	707-138B	Ex QANTAS ST Tracey Investment 9/75. RG 5/20/98 Jett Clipper Johnny Llc (John Travolta Qantas V-jet) Donated by John Travolta, 6/29/17, to be shipped to Australia for display
19104	N7095	5/21/66	707-327C	Braniff BT new. LSD TMA 4/14/71. BT: TMA 7/4/80. STD seen at BEY 2/24/02
19105	N7096	5/28/66	707-327C	Braniff BT new. LSD TMA 3/21/71. BT: TMA 7/9/80. STD 2/02 BEY all white. Derelict
19106	N7097	6/18/66	707-327C	Braniff BT new. ST Varig 11/24/71. On final approach to Rio, Varig jet nosed down and crashed into the sea. 6/9/73. OB:4 FT:2
19107	N7098	6/29/66	707-327C	Braniff BT new. LSD Transavia 6/18/70. BT TMA Cargo 7/1/72. WO TMA During training the 707 stalled and crash at Beirut 7/23/79. OB:7 FT:7
19108	N7099	7/27/66	707-327C	Braniff BT new. LSD QANTAS 1/1/71. BT TMA 10/1/72. WO. TMA Cargo, heavy landing on main gear, caused major damage on the wing primary structure. EHAM 7/26/93.
19374	N453PA	3/1/78	707-321B	LSD From Pan American 3/78. RTD 5/78
19440	N7100	2/17/67	707-327C	Braniff BT new. LSD TMA 4/29//71. BT: TMA 7/5/80. WO 7/7/81 Beirut during shelling of the airport causing heavy damage on aircraft.
19529	N7102	10/10/67	707-327C	Braniff BT new. ST Malaysia-Singapore AL 3/1/71.
19530	N7103	10/18/67	707-327C	Braniff BT new. ST Malaysia-Singapore AL 4/23/71. WFU SCR 2/15/95.
19531	N7104	11/20/67	707-327C	Braniff BT new. LSD TMA 8/1/73. BT: TMA 9/5/80. WO 11/98

Boeing 720

18041	N7083	11/15/65	720-048	Braniff LSD from EL AL, in Dark Blue livery 11/65. RTD 5/66
18042	N7081	9/15/64	720-048	Ex EL AL ST Pacific Northern 6/20/66. BU Boeing Field 9/80
18043	N7082	1/11/65	720-048	Braniff LSD from Aer Lingus 1/11/65. RTD 5/24/66
18064	N7076	2/11/61	720-027	Braniff BT new. ST American Aviation 9/20/73.
18065	N7077	3/22/61	720-027	Braniff BT new. LSD to Aeroamerica 4/75. BU KPAE 1984
18077	N7224U	5/6/72	720-022	Braniff LSD from United AL with Blue/Light Blue livery. RTD 12/72
18154	N7078	8/9/61	720-027	Braniff BT new. ST Aeroamerica 6/75. To Boeing Company 1983. BU KPAE 1988
18423	N7079	5/10/62	720-027	Braniff BT new. ST Club International 10/73. WFU 4/80 & BU at Bournemouth 4/82
18581	N7080	8/22/63	720-027	Braniff BT new. ST Aeroamerica 10/9/73. BU KPAE 1/82, CNC 10/12/83

Boeing 727

18794	N305BN	6/8/71	727-78	Ex BWIA ST Transbrasil 8/76.
18795	N306BN	4/10/71	727-78	Ex BWIA ST LAB 3/7/75. At LAB, over 79,000 landings. The most ever on a 727. STD SLCB 3/05
18796	N307BN	3/28/71	727-78	Ex BWIA LSD Faucett 5/3/75 RTD 6/30/75. ST LADECO 1/6/76.
18897	N303BN	5/29/69	727-51C	Ex Piedmont LSD Dominicana 3/29/72. RTD 5/15/75. ST Faucett 7/29/76.
19008	N310BN	5/16/79	727-30C	BT Evergreen LSD by Braniff [Panagra Green livery] 5/16/79. RTD 4/82
19012	N311BN	5/31/79	727-30C	BT Evergreen LSD by Braniff 5/31/79 [Dark brown livery] RTD 4/82
19109	N7270	5/26/66	727-27C	Braniff BT new. LSD American Flyers 11/1/67. RTD 5/1/68. ST Transbrasil 8/28/75.

19110	N7271	6/24/66	727-27C	Braniff BT new. ST Transbrasil 7/77.
19111	N7272	8/14/66	727-27C	Braniff BT new. ST Transbrasil 11/28/75. WO Poor weather conditions, and pilot error as aircraft was on high speed on approach, causing plane to crash on a hill at Florianopolis Brazil 4/12/80. OB:58 FT:55
19112	N7273	8/26/66	727-27C	Braniff BT new. ST Transbrasil 10/2/75. STD KMZJ 6/92
19113	N7274	9/14/66	727-27C	Braniff BT new. ST Transbrasil 12/22/76. STD KMZJ 4/92
19114	N7275	9/12/66	727-27C	Braniff BT new. ST UPS 3/11/82. CNC 8/23/06
19115	N7276	10/29/66	727-27C	Braniff BT new. LSD Faucett 6/75 RT 9/75.
19116	N7277	11/4/66	727-27C	Braniff BT new. LSD Transbrasil 4/26/77 RTD 11/77.
19117	N7278	2/28/67	727-27C	Braniff BT new. ST UPS 10/22/81.
19118	N7279	3/1/67	727-27C	Braniff BT new. ST UPS 10/5/81. STD KROW 2003.
19119	N7280	4/7/67	727-27C	Braniff BT new. ST UPS 10/24/81.
19120	N7281	4/17/67	727-27C	Braniff BT new. ST Air Guinee 5/22/80. LSD by First Aid. Retired CYRP. CNC 8/28/03
19242	N302BN	4/15/69	727-11	Braniff LSD from Wardair 4/15/69 RTD 3/26/70.
19243	N7282	5/20/66	727-162	Braniff BT new. ST Transbrasil 9/23/77. WFU Lima. BU 1995
19244	N7284	11/24/66	727-62C	Braniff BT new. ST UPS 10/20/81.
19245	N7286	12/12/66	727-62C	Braniff BT new. ST UPS 10/5/81.
19246	N7287	6/8/67	727-27C	Braniff BT new. ST UPS 10/28/81
19304	N2969G	3/15/69	727-193	Braniff LSD from Air West 3/15/69 RTD 2/27/70
19305	N2979G	3/1/69	727-193	Braniff LSD from Air West 3/1/69 RTD 3/6/70
19391	N297BN	12/6/69	727-191	Braniff LSD from Ledbetter 12/6/69 RTD 12/28/81
19392	N298BN	12/3/69	727-191	Braniff LSD from Ledbetter 12/3/69 RTD 1/6/82
19393	N299BN	4/17/69	727-191	Ex Frontier AL. LSD Transbrasil 3/14/80 BT Transbrasil 5/82.
19394	N300BN	11/5/69	727-191	Ex Frontier AL. ST Burlington Northern 1/4/82.
19395	N301BN	11/16/69	727-191	Braniff LSD from Ledbetter 11/16/69 RTD 1/6/82
19497	N7288	6/30/67	727-27C	Braniff BT new. ST Transbrasil 6/6/80.
19499	N7289	8/10/67	727-27	Braniff BT new. LSD Transbrasil 2/8/80 RTD 8/82.
19500	N7290	8/14/67	727-27	Braniff BT new. ST Transbrasil 9/29/78. WFU 12/9/04
19501	N7292	8/29/67	727-27	Braniff BT new. ST Transbrasil 12/10/79.
19506	N692WA	4/6/72	727-173C	Braniff LSD from World AW 4/6/172 RTD 6/7/73
19532	N7295	10/5/67	727-27C	Braniff BT new. ST Ladeco (Chile) 8/78
19533	N7296	10/20/67	727-27C	Braniff BT new. ST UPS 1/15/82. STD Lakland AFB TX CNC 8/23/06
19534	N7293	9/6/67	727-27	Braniff BT new. ST Alaska AL 5/17/76. BU 11/04
19535	N7294	8/31/67	727-27	Braniff BT new. ST Forbes Magazine 6/29/83.
19618	N530EJ	5/8/71	727-155C	Braniff LSD from Executive Jet Aviation 5/8/71 RTD 4/10/72.
19620	N898PC	2/1/69	727-193	Braniff LSD from Air West 2/1/69 RTD 2/5/70
19808	N309BN	3/21/72	727-172C	Ex Airlift International ST UPS 5/26/82.
19827	N12827	6/30/71	727-185C	Braniff LSD from Universal AL 6/30/71. Braniff BT 12/2/71.
19859	N1727T	10/22/72	727-171C	Braniff LSD from TIA 10/22/72 RTD 1/74.
19860	N1728T	27/11/72	727-171C	Braniff LSD from TIA 27/11/72 RTD 5/7/74
19991	N406BN	3/1/72	727-291	Ex Frontier AL LSD Southwest AL 3/1/79 RTD 1/14/80.
19992	N407BN	4/6/72	727-291	Ex Frontier AL ST Air One 12/28/83. BU KMIA 1998
19993	N408BN	3/13/72	727-291	Ex Frontier AL. This aircraft had the Calder Red-blue-white colors. WFU KOPF & was blown up in the movie "Bad Boys" 9/93.
20162	N7279F	4/24/69	727-214	Frontier AL LSD by Braniff 7/19/71, then BT by Braniff 12/1/71.
20217	N304BN	5/29/69	727-116	Braniff LSD from Boeing 5/29/69 RTD 12/19/69.
20302	N404BN	11/30/71	727-2B7	Ex Allegheny AL ST Northwest AL 6/15/83. BU TUS 9/93
20303	N405BN	11/25/71	727-2B7	Ex Allegheny AL ST Northwest AL 6/15/83.
20392	N401BN	6/30/70	727-227	Braniff BT new. ST Northwest AL 10/28/82.
20393	N402BN	6/30/70	727-227	Braniff BT new. ST Northwest AL 10/28/82. SCR 3/17
20394	N403BN	6/30/70	727-227	Braniff BT new. ST Avensa 2/20/81. BU for parts 2/98
20608	N410BN	7/3/72	727-227	Braniff BT new. ST American AL 3/9/81.
20609	N411BN	7/11/72	727-227	Braniff BT new. ST Europe Aero Service 4/15/80. STD at Perpignan France
20610	N412BN	7/19/72	727-227	Braniff BT new. ST American AL 12/24/80. BU KMHV 10/1/01.

20611	N413BN	8/1/72	727-227	Braniff BT new. ST American AL 1/9/81. CNC 3/26/13
20612	N414BN	3/13/73	727-227	Braniff BT new. ST American AL 2/20/81. WFU 10/7/01.
20613	N415BN	3/14/73	727-227	Braniff BT new. ST American AL 3/17/81. STD at KSBD 8/12/14
20729	N416BN	7/10/73	727-227	Braniff BT new. ST American AL 12/11/80.
20730	N417BN	7/12/73	727-227	Braniff BT new. ST American AL 3/25/81.
20731	N418BN	7/16/73	727-227	Braniff BT new. ST American AL 1/16/81.
20732	N419BN	8/10/73	727-227	Braniff BT new. ST American AL 11/17/80. WFU 2003
20733	N420BN	8/16/73	727-227	Braniff BT new. ST American AL 11/26/80.
20734	N421BN	8/21/73	727-227	Braniff BT new. ST American AL 1/11/81. WFU & STD Mojave 4/16/03
20735	N422BN	9/18/73	727-227	Braniff BT new. ST American AL 2/2/81. WFU. STD 2/15/08
20736	N423BN	9/24/73	727-227	Braniff BT new. ST American AL 1/11/81. STD CGK, Jakarta, Indonesia 12/6/05. BU
20737	N424BN	9/28/73	727-227	Braniff BT new. ST American AL 3/2/81. WFU & STD KROW
20738	N425BN	10/5/73	727-227	Braniff BT new. ST American AL 3/31/81. WFU Cargojet 1/18/08
20772	N426BN	10/23/73	727-227	Braniff BT new. ST PeopleExpress 2/15/84.
20773	N427BN	10/25/73	727-227	Braniff BT new. ST PeopleExpress 8/8/83. STD Latacunga Ecuador 12/31/02
20774	N428BN	12/14/73	727-227	Braniff BT new. ST PeopleExpress 6/10/83.WFU 2007 KIGM.
20775	N429BN	1/4/74	727-227	Braniff BT new. ST PeopleExpress 11/3/83. WFU SCR Kinshasa 4/07
20837	N430BN	2/27/74	727-227	Braniff BT new. ST PeopleExpress 1/30/84. WFU 12/99.
20838	N431BN	3/6/74	727-227	Braniff BT new. ST PeopleExpress 7/8/83. Converted to 727-227(F) 12/95. WFU 4/21/17
20839	N432BN	4/19/74	727-227	Braniff BT new. ST PeopleExpress 7/8/83. BU 6/16/16 at CYHM.
20840	N433BN	5/9/74	727-227	Braniff BT new. ST PeopleExpress 8/9/83
21041	N434BN	2/4/75	727-227	Braniff BT new. ST PeopleExpress 9/30/83. BU 5/05
21042	N435BN	2/20/75	727-227	Braniff BT new. ST PeopleExpress 10/27/83.
21043	N436BN	3/12/75	727-227	Braniff BT new. ST PeopleExpress 8/27/83.
21044	N437BN	5/14/75	727-227	Braniff BT new. ST PeopleExpress 6/15/83. WFU 2010
21045	N438BN	5/14/75	727-227	Braniff BT new. ST PeopleExpress 9/14/83.
21118	N439BN	10/1/75	727-227	Braniff BT new. ST PeopleExpress 8/26/83. BU at CYHM 6/16
21119	N440BN	12/2/75	727-227	Braniff BT new. ST PeopleExpress 7/12/83.
21242	N441BN	4/22/76	727-227	Braniff BT new. ST PeopleExpress 8/29/83. STD 12/2000 at SBGR Sao Paulo International Airport
21243	N442BN	4/23/76	727-227	Braniff BT new. ST PeopleExpress 10/14/83.
21244	N443BN	4/29/76	727-227	Braniff BT new. ST PeopleExpress 9/27/83.
21245	N444BN	5/4/76	727-227	Braniff BT new. ST PeopleExpress 5/2/84.
21246	N445BN	7/30/76	727-227	Braniff BT new. ST PeopleExpress 11/22/83. STD Lima Peru 2009
21247	N446BN	7/27/76	727-227	Braniff BT new. ST Continental AL 5/30/85. STD at SNBO, Brazil.
21248	N447BN	10/29/76	727-227	Braniff BT new. ST Continental AL 5/30/85. WFU 6/15/10
21249	N448BN	10/28/76	727-227	Braniff BT new. ST Continental AL 5/30/85. STD 6/7/17 Philippines
21363	N449BN	4/20/77	727-227	Braniff BT new. ST Continental AL 5/30/85. STD GRU Brazil 5/30/07
21364	N450BN	5/19/77	727-227	Braniff BT new. ST Alaska AL 4/30/85. WFU Lagos, Nigeria 8/13
21365	N451BN	7/8/77	727-227	Braniff BT new. ST Alaska AL 4/30/85. BU KVQQ . CNC 9/20/11
21366	N452BN	7/7/77	727-227	Braniff BT new. ST Continental AL 5/30/85. WFU CNC 4./29/03
21394	N453BN	7/27/77	727-227	Braniff BT new. With Braniff till end of operations 5/13/82. WFU CNC 1/14/10
21395	N454BN	8/1/77	727-227	Braniff BT new. With Braniff till end of operations 5/13/82.

21461	N455BN	4/27/78	727-227	Braniff BT new. With Braniff till end of operations 5/13/82. WFU 8/10/15
21462	N456BN	5/10/78	727-227	Braniff BT new. With Braniff till end of operations 5/13/82. WFU 10/23/14
21463	N457BN	6/13/78	727-227	Braniff BT new. With Braniff till end of operations 5/13/82. WFU & STD EGHL. CNC 10/18/18
21464	N458BN	6/15/78	727-227	Braniff BT new. With Braniff till end of operations 5/13/82. STD Port of Spain 7/6/15
21465	N459BN	7/17/78	727-227	Braniff BT new. With Braniff till end of operations 5/13/82. WFU 8/8/12. Donated to Kansas City Airport Fire Dept. KMCI.
21466	N460BN	8/9/78	727-227	Braniff BT new. With Braniff till end of operations 5/13/82. WFU CNC 03/25/13. Stored at Victorville CA
21488	N461BN	9/21/78	727-227	Braniff BT new. With Braniff till end of operations 5/13/82. WFU Donated to Albuquerque Airport. CNC 8/27/13
21489	N462BN	9/27/78	727-227	Braniff BT new. With Braniff till end of operations 5/13/82. WFU 7/31/12 Donated to Pima Community College KTUS.
21490	N463BN	10/11/78	727-227	Braniff BT new. With Braniff till end of operations 5/13/82. WFU CNC 3/8/13
21491	N464BN	10/26/78	727-227	Braniff BT new. With Braniff till end of operations 5/13/82. WFU
21492	N465BN	2/7/79	727-227	Braniff BT new. With Braniff till end of operations 5/13/82. WFU 7/1/13
21493	N466BN	2/9/79	727-227	Braniff BT new. With Braniff till end of operations 5/13/82. WFU 10/2314. STD 10/16
21529	N467BN	2/19/79	727-227	Braniff BT new. With Braniff till end of operations 5/13/82. WFU CNC 8/27/13
21530	N468BN	2/22/79	727-227	Braniff BT new. With Braniff till end of operations 5/13/82. WFU Donated to University of Alaska Anchorage for ground training.
21531	N469BN	3/2/79	727-227	Braniff BT new. With Braniff till end of operations 5/13/82. WFU To Everett Community College Aviation School (KPAE) 7/25/13.
21532	N470BN	3/10/79	727-227	Braniff BT new. With Braniff till end of operations 5/13/82. WFU 1/31/13. Donated to Rocky Mtn. Emergency Service Training, Helena, MT.
21669	N471BN	5/25/79	727-227	Braniff BT new. With Braniff till end of operations 5/13/82. WFU Donated to Mid America Air Museum Foundation. CNC 8/21/14
21670	N472BN	5/31/79	727-227	Braniff BT new. With Braniff till end of operations 5/13/82. WFU Donated to Grand Forks Airport Fire Department 8/28/07. CNC 11/14/07
21996	N473BN	1/22/80	727-227	Braniff BT new. ST US Air 12/13/83. WFU STD 2/13/17
21997	N474BN	1/29/80	727-227	Braniff BT new. With Braniff till end of operations 5/13/82. WFU STD at Opa-Locka 12/18/03 BU
21998	N475BN	2/7/80	727-227	Braniff BT new. With Braniff till end of operations 5/13/82. WFU STD 5/22/13
21999	N476BN	2/24/80	727-227	Braniff BT new. With Braniff till end of operations 5/13/82. WFU STD 5/22/13
22000	N477BN	2/21/80	727-227	Braniff BT new. With Braniff till end of operations 5/13/82. WFU Boca Raton FL. CNC 02/16/05
22001	N478BN	2/28/80	727-227	Braniff BT new. With Braniff till end of operations 5/13/82. WFU STD KIGM 5/11/09
22002	N479BN	6/6/80	727-227	Braniff BT new. With Braniff till end of operations 5/13/82. WFU To KSFB RG expired 9/30/15

Boeing 747

19746	N610BN	11/29/78	747-130	Ex Lufthansa. LSD by Braniff from ITEL Corp. 11/29/78 RTD 10/23/81 to GATX Leasing 9/30/82.
20105	N9666	3/1/78	747-123	Braniff LSD from American AL 3/1/78 RTD to Citicorp Industrial 6/7/84
20106	N9667	10/29/78	747-123	Braniff LSD from American AL 10/29/1978 RTD 11/16/78

20207	N601BN	1/5/71	747-127 SCD	Braniff BT new. ST Polaris 6/82. WFU 3/21/92.
20527	N611BN	5/11/79	747-230B	Ex Lufthansa. ST British AW 4/14/88. WFU KROW 11/07
20653	N749WA	4/10/79	747-273C	Braniff LSD from World AW 4/10/1979 RTD 6/9/81. Converted to waterbomber 5/04. STD at KMZJ 2/25/12
20927	N620BN	11/15/78	747-217B	Braniff LSD from CP Air 11/15/78 RTD 12/7/78
21682	N602BN	5/31/79	747-227B	Braniff BT new. ST PeopleExpress 9/84. WFU
21785	N603BN	10/30/79	747SP-27	Braniff BT new. ST Boeing Corporation 1/23/81. To Oman gvt. Oman Royal Flight 7/2/84
21786	N604BN	4/23/80	747SP-27	Braniff BT new. ST Aerolineas Argentinas 9/12/80. ST Yemen Rvt. WO Aircraft was destroyed in rival clashes during civil war. Plane was sprayed with bullets. 3/19/15
21992	N606BN	5/30/80	747SP-27	Braniff BT new. ST Pan American 9/23/83. RG to NASA - Armstrong Flight Research Center Edwards CA 1/10/17

Convair 240

9	N90655	8/15/52	240-0	Ex Mid-Continent Airlines. ST Linee Aeree Italiane 3/7/53. BU
55	N90663	8/15/52	240-2	Ex Mid-Continent Airlines. ST Linee Aeree Italiane 3/20/53. WO Air Jordan Crashed 1/22/59 on approach to Amman Jordan.OB:15 FT:10.
71	N90667	8/15/52	240-2	Ex Mid-Continent Airlines. ST Linee Aeree Italiane 12/31/52. Comisión Federal de Electricidad, crashed 1/25/70 near Poza Rica-Tajin Airport. Mexico OB: 20 FAT:19. WO
84	N90669	8/15/52	240-2	Ex Pan American. ST Linee Aeree Italiane 4/9/53. WFU at Perryville Missouri 1971. BU

Convair 340/440

6	N3406	7/26/52	340-32	Braniff BT new. ST Caribbean Atlantic Airlines 4/30/64. WFU CNC 6/24/1976
20	N3407	10/15/52	340-32	Braniff BT new. ST Caribbean Atlantic Airlines 7/30/65. WFU 1995. STD KROW
21	N3408	10/29/52	340-32	Braniff BT new. ST Caribbean Atlantic Airlines 3/66. WFU/WO 1/23/67
26	N3409	11/29/52	340-32	Braniff BT new. ST Caribbean Atlantic Airlines 4/12/62.
27	N3410	11/20/52	340-32	Braniff BT new. ST Caribbean Atlantic Airlines 6/1/62.
31	N3411	12/16/52	340-32	Braniff BT new. ST Caribbean Atlantic Airlines 3/28/62. CV to 640 6/66. WFU DBR Both engines quit at 14,000ft. Forced belly landing at Bartlesville, OK. 10/28/87. No fatalities
32	N3412	12/17/52	340-32	Braniff BT new. ST Caribbean Atlantic Airlines 9/4/62. WFU
37	N3413	1/19/53	340-32	Braniff BT new. WFU 1967. ST Charlotte Aircraft Corporation 6/1/1971.
45	N3414	1/29/53	340-32	Braniff BT new. ST Allegheny Airlines 3/62. CV to 440 7/67. WO DBR Overran runway while on aborted takeoff Harrisburg, PA. Reverse thrust did not work due to failure of the electrical system 11/29/66. No fatalities
46	N3415	2/6/53	340-32	Braniff BT new. ST Allegheny Airlines 6/20/62. WFU Used as artificial reef at Sunny Isles Beach, Opa Locka FL.
47	N3416	2/10/53	340-32	Braniff BT new. WFU 1967. ST Charlotte Aircraft Corporation 6/1/1971. WFU STD KOPF 2/7/09
48	N3417	2/10/53	340-32	Braniff BT new. ST Caribbean Atlantic Airlines 10/30/64.
55	N3430	4/2/55	340-42	Ex Philippine AL. ST Executive Aircraft Services 12/6/61. CV to 440 1963. CV to 580 6/69. WFU1999. STD at CCS - Simón Bolívar Int. Airport.
61	N3418	3/1/53	340-32	Braniff BT new. ST North Central Airlines 8/67. CV to 580 2/68. WFU 1999
62	N3419	4/53	340-32	Braniff BT new. ST Allegheny Airlines 10/31/66. CV to 580 2/67. WFU SCR Bogota 7/13

64	N3420	4/9/53	340-32	Braniff BT new. ST Allegheny Airlines 11/66. CV to 640 1/66. WFU STD at Cleveland 6/2/15
65	N3421	4/9/53	340-32	Braniff BT new. ST Allegheny Airlines 11/66. CV to 580 2/67. WFU 05/29/08
71	N3422	5/29/53	340-32	Braniff BT new. WO Braniff Flight no.BN560, crashed on approach to Chicago-Midway Airport 7/17/55. The aircraft clipped a sign and crashed through the airport boundary fence onto the airport. Momentary disorientation by the pilots was a probable cause. OB:43 FT:22
77	N3423	5/22/53	340-32	Braniff BT new. ST North Central Airlines 10/17/67. CV to 580 3/68. WFU
78	N3424	5/22/53	340-32	Braniff BT new. ST North Central Airlines 7/12/66. CV to 580 3/68. To Conair, CV to Firefighter 1/29/01
79	N3431	4/2/55	340-42	Ex Philippine AL. ST Pacific Airmotive Corp. 8/9/61. CV to 580 12/61. WO at Maramba, Zambia. Emergency landing after takeoff. Aircraft was DBR 11/18/99
80	N3425	6/30/53	340-32	Braniff BT new. ST Frontier Al 12/27/65. CV to 580 4/66. WO Conair Firefighter. After landing aircraft suffered directional control problems and veered off runway to the right. It then hit a drainage ditch, lost its nose gear and came to rest in a grassy area 5/5/16. No fatalities.
81	N3426	7/10/53	340-32	Braniff BT new. ST Allegheny Airlines 3/28/66. CV to 580 4/66. WFU
90	N3427	7/7/53	340-32	Braniff BT new. ST John R. Morey 9/67. STD McAllen TX KMFE
99	N3428	8/7/53	340-32	Braniff BT new. ST Fred Olsen Flyselskap (Norway) 9/16/68. WFU.
109	N3429	10/28/53	340-32	Braniff BT new. ST North Central Airlines 7/1/66. CV to 580 3/67. WO Aircraft veered off the runway on landing at CYAH Quebec, Canada 9/27/2000. No fatalities.
110	N3430	10/28/53	340-32	Braniff BT new. ST Ansett AL (Australia) 6/10/64. CV TO 440 1960. CV to 580 4/66. WFU 9/9/99
375	N3432	11/20/56	440-32	Braniff BT new. ST Allegheny Airlines 8/11/66. CV to 580 8/67. BU 9/8/11
376	N3433	11/20/56	440-32	Braniff BT new. ST Allegheny Airlines 9/8/66. CV to 580 11/67. WFU STD at CYUL Montréal-Trudeau International Airport 8/2000
384	N3434	12/17/56	440-32	Braniff BT new. ST Allegheny Airlines 6/6/62. CV to 580 11/67. WO Due to very poor weather conditions, Pilot passed below the glide slope and continued the approach until the airplane struck cottages and crashed in flames at New Haven Connecticut, 6/7/71. OB:31 FT:28
385	N3435	12/28/56	440-32	Braniff BT new. ST Allegheny Airlines 4/16/62. CV to 580 8/67. WFY STD at PMR Palmerston North Airport 10/25/07
386	N3436	1/3/57	440-32	Braniff BT new. ST Allegheny Airlines 5/13/62. CV to 580 8/67. WO Aircraft clipped treetops 4.7nm short of the airport and cut a swath through trees bordering a fairway of the Pine Acres Golf Course. The aircraft came to rest upside down at Bradford, PA. 1/6/69. OB:28 FT:11
453	N3437	12/2/57	440-0	Braniff BT new. ST Allegheny Airlines 10/5/66. CV to 580 30/12/66. WFU 1980. BU KYIP 10/81

Curtiss C-46 Commando

22265	N1386N	11/55	C-46D-20-CU	Ex Civil Air Transport (Taiwan). ST Air America 10/1/63. Crashed on take-off after engine failure 11/25/68 Savannakhet Laos. OB:28 FT: 26. WO
33641	N1383N	11/55	C-46D-15-CU	Ex Civil Air Transport. ST Air America 10/10/63. ST Oasis Philippines WFU 1980. BU

Douglas DC-3

1963	NC17338	8/15/52	DC-3-208	Ex Mid-Continent Airlines. Braniff MGD with MCA 8/16/52. ST Southeastern Aviation 9/20/1958. Air BVI N4425N. Destroyed by Hurricane Hugo 9/18/89. WO
2179	NC21773	12/15/39	DC-3-134	Braniff BT new. ST Southern Airways 6/15/1959. BTD Philippine govt.1978. STD
2180	NC21774	12/28/39	DC-3-134	Braniff BT new. ST United Transports Inc 9/50.
2181	NC21775	12/30/39	DC-3-134	Braniff BT new. ST LAB - Lloyd Aereo Boliviano /6/3/53. LAB C-600. Crashed onto Rodeo Pampa mountain Bolivia, 11/3/53. OB:28 FT:28
2182	NC21776	12/29/39	DC-3-134	Braniff BT new. ST LAB - Lloyd Aereo Boliviano /6/3/53. STD Trinidad- Santa Cruz 4/18/91
2210	N25670	8/15/52	DC-3A	Ex Mid-Continent Airlines. Braniff MGD with MCA 8/16/52. SCR 1982
2212	N25672	7/19/40	DC-3-277B	Ex Mid-Continent Airlines. Braniff MGD with MCA 8/16/52. ST Lake Central Airlines 1/30/1961 WFU at Bali-Denpasar Airport in 1972.BU for spares 2/78
2216	N25685	8/15/52	DC-3A	Ex Mid-Continent Airlines. Braniff MGD with MCA 8/16/52. ST Trans Texas Airways 3/15/1959. At Cape Cod 8/5/15
2239	NC25666	6/20/40	DC-3-314A	Braniff BT new. ST Trans Texas Airways 4/53. RG regular visitor to airshows
2240	NC25667	6/20/40	DC-3-314A	Braniff BT new. ST Southern Airways 6/15/59. STD 1978 Philippines
2241	NC25669	6/20/40	DC-3-314A	Braniff BT new. ST Trans Texas Airways 6/53. TTA WFU 2/80.
2242	NC25669	6/20/40	DC-3-314A	Braniff BT new. ST Cordova Airlines 8/52. STD
2266	NC25693	0/0/44	DC-3-313A	Ex Pennsylvania-Central Airlines. ST PLUNA 2/3/48. WFU & BU 8/67
3275	N41831	8/15/52	DC-3-363	Ex Mid-Continent Airlines. Braniff MGD with MCA 8/16/52. ST Lake Central Airlines 5/1/53. WFU 9/81. BU
3280	42-65584	03/05/43	DC-3-357	To USAAF Jul 22, 1942. Operated by Braniff 5/3/43, and United 6/18/43. CNC poor condition 1980
4106	NC28362	6/20/41	DC-3-314B	Braniff BT new. ST Southern Airways 1/4/60. WFU Philippine AL 1973.
4107	NC28363	6/20/41	DC-3-314B	Braniff BT new. ST Southern Airways 12/50. WFU Manila 7/78. BU
4108	NC28364	6/20/41	DC-3-314B	Braniff BT new. ST Trans Texas Airways 5/31/53. WO Crashed at Ciudad Camargo, Mexico, while landing. Caught fire. FT:1
4125	NC33646	10/5/42	DC-3A-197E	Ex United AL. ST United AL. Ground accident 2/24/47 at Philadelphia, PA. Aircraft WO.
4127	N33326	8/15/52	DC-3-395	Ex Mid-Continent Airlines. Braniff MGD with MCA 8/16/52. ST W.F. Long 10/5/59. RG CNC 10/3/12. BU
4128	N33327	8/15/52	DC-3-395	Ex Mid-Continent Airlines. Braniff MGD with MCA 8/16/52. ST Trans Texas Airways 4/14/59..
4390	N61351	8/15/52	DC-3	Ex Mid-Continent Airlines. Braniff MGD with MCA 8/16/52. ST Tusco Corporation 1/21/60. Derelict at Denpasar Bali 1972. BU
4535	N61350	8/15/52	DC-3	Ex Mid-Continent Airlines. ST Air Oasis 10/25/59. Hawthorne Nevada Airlines. Both engines out, emergency landing attempted & crashed at Tonopah, NV. 8/19/64. No fatalities.
4630	N61451	8/15/52	DC-3	Ex Mid-Continent Airlines. WO Crashed in thunderstorm near Mason City, IA on August 22, 1954. OB:19 FT:12
4736	NC65350		DC-3	Ex USAAF. ST Indamer Company. Destroyed in a hangar fire on delivery at Miami 8//6/47
4822	N34951	8/15/52	DC-3	Ex USAAF. ST Southern Airways 1/14/53. Crashed 1/21/65. Left main gear collapsed while landing. Aircraft DBR. No fatalities.
4827	N34951	8/15/52	DC-3	Ex Mid-Continent Airlines. Braniff MGD with MCA 8/16/52. ST Trans Texas Airways 11/18/58.

4842	N33312	8/15/52	DC-3	Ex Mid-Continent Airlines. Braniff MGD with MCA 8/16/52. ST Lake Central Airlines 5/20/53. RG CNC 6/12/12
4944	N34952	8/15/52	DC-3	Ex Mid-Continent Airlines. Braniff MGD with MCA 8/16/52. ST Trans Texas Airways 12/59. STD
4964	N45335	8/15/52	DC-3A	Ex Mid-Continent Airlines. Braniff MGD with MCA 8/16/52. ST Lake Central Airlines 4/14/53. WO 11/1/63 caught fire inside a hangar at Indianapolis, IN. RG CNC 12/2/63
4965	N21914	8/15/52	DC-3A	Ex Mid-Continent Airlines. Braniff MGD with MCA 8/16/52. BU
4976	NC45340	00/11/45	C-53C	Ex USAAF. ST Southern Airways 12/31/52. RG CNC 5/31/13. BU
7349	N28508	8/15/52	DC-3A	Ex Mid-Continent Airlines. Braniff MGD with MCA 8/16/52. ST Trans Texas Airways 11/18/59. RG CNC 11/7/68. BU
7356	NC33346		DC-3A	Ex USAAF. ST Southern Airways 1954. RG CNC 2/21/13. BU
7399	N34950	8/16/52	DC-3A	Ex Mid-Continent Airlines. Braniff MGD with MCA 8/16/52. ST Lake Central Airlines 3/15/59. BU Conroe-North Houston Regional KCXO 1975
7403	NC49556	0/0/45	DC-3A	Ex USAAF. ST Southern Airways. WFU.
10035	N28679	8/16/52	DC-3C	Ex Mid-Continent Airlines. Braniff MGD with MCA 8/16/52. ST Stewart Air Services 8/21/58. STD
11644	N18667	8/15/52	DC-3A	Ex Mid-Continent Airlines. Braniff MGD with MCA 8/16/52. ST Lake Central Airlines 2/15/59. ST Indonesia 1972.
11666	N15201	8/15/52	DC-3A	Ex Mid-Continent Airlines. Braniff MGD with MCA 8/16/52. ST Southern Airways 6/19/1960. Philippine Airlines STD 1979
11682	N17888	8/15/52	DC-3A	Ex Mid-Continent Airlines. Braniff MGD with MCA 8/16/52. ST Tusco 5/20/1960. WFU 9/7/69.
11877	NC59748	11/13/45	DC-3C	Ex USAAF. ST Transair (Canada) 4/10/1959. BU
19225	NC65351		DC-3C	Ex USAAF. ST West Coast AL 1946. BU
19238	NC65378		DC-3C	Ex USAAF. ST Chas E. Matthews 10/16/47. Aerolíneas TOA - shortly after takeoff plane stalled & crashed at San Vicente del Caguán, Neiva Colombia 10/17/71. OB:21 FT:19
20196	NC59749	0/0/1945	DC-3C	Ex USAAF. ST West Productions 3/49.
42967	N34971	8/15/52	DC-3D	Ex Mid-Continent Airlines. Braniff MGD with MCA 8/16/52. ST Southern Airways 2/19/1960. Filipinas Orient Airways. Destroyed by typhoon Patsy in Philippines. 11/19/70. No Fatalities.

Douglas DC-4/C-54

10282	N65144	00/11/45	C-54A-5-DC	Ex USAAF. ST Los Angeles Air Service 6/54. WFU & STD 1963. BU
10336	N65143	00.11.45	C-54A-10-DC	Ex USAAF. Braniff DBR 3/26/52. Fire in the No. 3 engine caused a necessary landing on runway. Plane crashed through airport boundary onto road. No Fatalities
10338	N65142	00/11/45	C-54A-10-DC	Ex USAAF. ST Resort Airlines 11/54. Cie Air Transport F-BMHU - After takeoff failed to gain height & Crashed onto bridge. 3/8/67. OB: 6 FT:13 (with 7 on ground)
10360	N65145	00/11/45	C-54A-15-DC	Ex USAAF. ST Los Angeles Air Service 5/54. Wheeler airlines WO. Crashed near St. Cleophis Quebec 11/4/59. OB:5 FT:5
10424	N65141	00/11/45	C-54A-15-DC	Ex USAAF. ST Nevada Aero Traders 7/54. TMA (Lebanon) OD-AEB crashed in Ghazni Afghanistan 12/12/63. OB:3 FT:3
10528	N59952	00.00.47	C-54B-1-DC	Ex Douglas Aircraft Corp. ST World Airways 6/53. WFU Lydd Airport 3/67 BU 7/70
10539	N86575	00.11.45	C-54B-1-DC	Ex USAAF. ST Grimley Engineering Corp. 11/54, WFU N'Djamena 1980
18339	N88709	00.11.45	C-54B-5-DO	Ex USAAF. ST Northwest Airlines 10/50. Channel Air Bridge G-ARSF crashed 800 feet short of runway Rotterdam Airport 12/28/62. OB:18 FT:1
18353	N88818	00.11.45	C-54B-10-DO	Ex USAAF. ST Chicago & Southern Airlines 1/29/49. SCR.
18383	N86573	00.11.45	C-54B-15-DO	Ex USAAF. LSD Northwest Airlines 1946. WFU Vancouver3/69. BU

Douglas DC-6

43105	N90881	8/23/47	DC-6	Braniff BT new. ST La Jolla Associates 7/23/65. WFU & STD Mexico City 10/71
43106	N90882	8/10/47	DC-6	Braniff BT new. ST La Jolla Associates 7/30/65. BU at Lima, Peru
43107	N90883	10/25/47	DC-6	Braniff BT new. ST La Jolla Associates 1/24/65.
43108	N90884	10/28/47	DC-6	Braniff BT new. ST La Jolla Associates 2/4/66. BU 5/81
43109	N90885	11/3/47	DC-6	Braniff BT new. ST La Jolla Associates 2/4/66. WFU 1979
43110	N90886	11/14/47	DC-6	Braniff BT new. ST Mexicana de Aviación 12/16/65. DBR 8/25/70 Bahia de Tortugas Mexico.
43293	N6887	8/23/51	DC-6	Braniff BT new. ST La Jolla Associates 12/30/65. BU 1969 Ensenada, Mexico
43294	N6888	9/30/51	DC-6	Braniff BT new. ST La Jolla Associates 1967. 1970 WFU & STD Lima Peru.
43295	N6889	11/2/51	DC-6	Braniff BT new. ST Aerocarga 8/30/66. BU 1975 Ensenada Mexico
43839	N90776	5/16/53	DC-6A	Ex Los Angeles Air Service. ST Aerocarga 1967. BU 1981
45063	N91311	0/0/58	DC-6B	Ex Western AL. ST Lan Chile 10/6/65. WFU On display at Parque La Carolina, Quito Ecuador.

Douglas DC-7

44703	N51703	2/1/67	DC-7B	Ex Panagra. Braniff MGD with Panagra 2/1/67. ST TAN Honduras 1967. BU Miami 1970
44704	N51704	9/22/55	DC-7B	Ex Panagra. Braniff MGD with Panagra 2/1/67. ST TAN Honduras 4/13/67. BU Miami1970
45068	N5900	9/13/56	DC-7C	Braniff BT new. ST Air Travel Club Inc. 2/20/68. STD Burbank CA 1980 BU
45069	N9501	10/18/56	DC-7C	Braniff BT new. WFU ST Charlotte Aircraft Corp. 9/70. Sebring Florida BU 6/24/76
45070	N5902	10/24/56	DC-7C	Braniff BT new. WFU ST Charlotte Aircraft Corp. 9/70. STD Chandler AZ 1990. BU
45071	N5903	11/30/56	DC-7C	Braniff BT new. ST Club America Inc. 11/30/69. T & G Aviation. WO. 10/9/86 at Dakar, Senegal. Crashed after takeoff with engine fire. OB:4 FT:3
45072	N5904	4/13/57	DC-7C	Braniff BT new. WO 3/25/58 at Miami. No. 3 engine fire crashed in marsh. OB:24 FT:9
45073	N5905	5/10/57	DC-7C	Braniff BT new. WO. Destroyed in ground fire at KDAL Dallas. 11/14/61
45074	N5906	5/16/57	DC-7C	Braniff BT new. ST Charlotte Aircraft Corp. 9/70. WFU & STD 9/70 AZ. BU

Douglas DC-8

45274	N8274H	2/1/67	DC-8-31	Braniff MGD with Panagra 2/1/67 & received this DC-8. ST Capitol International AW 11/9/67. WFU STD Waco TX 3/83
45275	N8275H	2/1/67	DC-8-31	Braniff MGD with Panagra 2/1/67 & received this DC-8. ST United AL 9/5/67. CV to Freighter 6/77. WFU Miami 1/81, BU 12/83
45276	N8276H	2/1/67	DC-8-31	Braniff MGD with Panagra 2/1/67 & received this DC-8. Braniff RG N1801. ST Capitol International AW 11/9/67. BU at KMFR 6/82
45277	N8277H	2/1/67	DC-8-31	Braniff MGD with Panagra 2/1/67 & received this DC-8. Braniff RG N1802. WO Capitol Air: crashed on training touch & go landings with two engines working. Pilot had attempted a go-around when then aircraft lost height and crashed in flames short of runway threshold 4/28/68. No FT
45634	N811BN	10/24/73	DC-8-51	Ex National AL. BT Braniff 11/7/73. BU Miami 1/83

45635	N812BN	9/1/73	DC-8-51	Ex National AL. ST Hernando Gutierrez 2/15/80. SCR 2001
45642	N813BN	6/15/73	DC-8-51	Ex National AL. ST Sonico Inc 9/82. WFU BU12/82.
45644	N814BN	8/10/73	DC-8-51	Ex National AL. ST Sonico Inc 9/82. WFU BU 82
45815	N820E	7/28/77	DC-8-51	Braniff LSD from Delta AL 7/28/77. RTD 2/15/80
45858	N1509U	2/1/67	DC-8-55F	Braniff MGD with Panagra 2/1/67 & received this DC-8. ST Canadian Pacific AL 11/17/67. WFU Kinshasa, Zaire 5/12/09
45877	N821E	7/28/77	DC-8-51	Braniff LSD from Delta AL 7/28/77. RTD 2/22/80
45895	N1803	8/22/67	DC-8-62	Braniff received from Panagra. ST International Air Leases 11/83. WFU
45896	N1804	9/20/67	DC-8-62	Braniff received from Panagra. ST International Air Leases 11/83. WO Damaged on 12/13/02. Arrow Air to Singapore inbound from Yokota Japan, in heavy rain, overran runway & aircraft came to rest 300 meters past end of runway. CNC 11/8/13. No Fatalities
45899	N1805	9/29/67	DC-8-62	Braniff received from Panagra. ST International Air Leases 10/83. WFU.
45904	N1807	11/13/67	DC-8-62CF	Braniff received from Panagra. ST International Air Leases 11/83. WFU. BU at EBOS 4/2/13
45905	N810BN	1/12/77	DC-8-62	Ex SAS. ST International Air Leases 12/1/83. STD at KROW 1/28/05.
45909	N802BN	11/24/78	DC-8-62H	Ex Alitalia. ST International Air Leases 11/83. WFU STD at KMIA.
45911	N1806	12/19/67	DC-8-62	Braniff received from Panagra. ST International Air Leases 11/83. WFU.
46082	N801BN	7/1/79	DC-8-62H	Ex Alitalia. ST International Air Leases 11/83. BT NASA CV to DC-8-72 1986
46105	N1808E	10/23/69	DC-8-62	Braniff BT new. ST International Air Leases 6/83. damaged at Singapore when aircraft fell in drainage ditch 3/1/02. SCR
46107	N1809E	11/17/69	DC-8-62	Braniff BT new. ST International Air Leases 6/83. WO Surinam AW - Pilot error caused the aircraft to be flown below the published minimum altitudes during the approach and it consequently collided with a tree at Paramaribo, Surinam. OB:187 FT:176

Lockheed L-049 Constellation

2081	N2520B	8/5/55	L-049	Ex LAV Venezuela. ST Hubert M Covert 11/7/60 ST Preserved at Santa Cruz.
2082	N2521B	8/10/55	L-049	Ex LAV Venezuela. ST Hubert M Covert 3/24/61. WFU Miami 1967 BU 1968

Lockheed L-188 Electra

1004	N16816	10/12/66	L-188A	Ex Frederick B. Ayer & Associates. Braniff ST Frederick B. Ayer & Associates 10/14/67. BU Quito, Ecuador 1981
1040	N9701C	4/29/59	L-188A	Braniff BT new. RG to Crocker Citizens National Bank 3/28/68. RTD to Braniff 3/27/69. ST Frederick B. Ayer & Associates 5/26/70. WO 7/16/03 Cranbrook, Canada
1052	N9702C	5/6/59	L-188A	Braniff BT new. RG to Crocker Citizens National Bank 3/28/68. RTD to Braniff 3/27/69. ST Boeing Company 2/10/71. SELA Ecuador, after takeoff no. 2 engine was out, and with explosion and flames from no. 1 engine, caused aircraft to veer left and crash on 9/12/88. OB:7 FT:7
1067	N9703C	6/22/59	L-188A	Braniff BT new. RG to Crocker Citizens National Bank 3/28/68. RTD to Braniff 3/27/69. ST Boeing Company 11/9/70. WFU 1991 Argentina
1086	N9704C	8/28/59	L-188A	Braniff BT new. RG to Crocker Citizens National Bank 3/28/68. RTD to Braniff 3/27/69. ST Boeing Company 12/10/69. WO 12/24/71 Aircraft crashed at Puerto Inca Peru in severe turbulence, and lightning which ripped off both wings. OB; 92 FT:91.

1090	N9705C	9/18/59	L-188A	Braniff BT new. WO. Structural failure caused catastrophic disintegration of aircraft over Buffalo TX 9/29/59. OB:34 FT:34
1095	N9706C	1/10/59	L-188A	Braniff BT new. RG to Crocker Citizens National Bank 3/28/68. RTD to Braniff 3/27/69. ST Boeing Company 10/9/70. WFU Lima 1971. BU
1099	N9707C	10/17/59	L-188A	Braniff BT new. RG to Crocker Citizens National Bank 3/28/68. Braniff N9707C crashed at Dawson TX. Extremely bad weather and stressing of the aircraft frame caused the breakup and crash 3/5/68. OB:85 FT:85
1106	N9708C	11/17/59	L-188A	Braniff BT new. RG to Crocker Citizens National Bank 3/28/68. RTD to Braniff 3/27/69. ST Boeing Company 10/15/69. WO LANSA. Engine no.3 failed on takeoff, control was lost, and the airplane crashed onto a hill 4 km east of the airport at Lima, Peru 8/9/70. OB:102 FT:101
1114	N9709C	1/9/60	L-188A	Braniff BT new. RG to Crocker Citizens National Bank 3/28/68. RTD to Braniff 3/27/69. ST Frederick B. Ayer & Associates 7/10/70. BU 1997
1134	N9710C	5/10/62	L-188C	Capital Airlines NTU. ST Braniff then RG to Crocker Citizens National Bank 3/28/68. RTD to Braniff 3/27/69. ST Boeing Company 4/28/69. TAN Honduras 3/21/90 crashed on approach to Tegucigalpa in bad weather. Flew into the southern slopes of the Cerro Hula mountain. OB:3 FT:3

BAC/Aerospatiale Concorde - Braniff International had an agreement with both Air France and British Airways to operate the Concorde within the U.S for one year, from 1979 to 1980. Basically, Braniff was allowed by the two European carriers to use the Concorde on the Washington Dulles to Dallas run and back. The European airlines would then fly it to Europe and back. During the U.S. domestic part of the run, all the Concordes were given temporary U.S. "N" registrations. Also, they were all manned by Braniff crews. The following Concordes were the ones used by Braniff: N81AC, N94FA, N94AA, N94FB, N94AB, N94FC, N94AD, N94FD, N94AE. In total 9 Concordes were utilized over the year, out of a total 20 Concordes that were produced.

BIBLIOGRAPHY/RESOURCES

Printed books:

Bluffield, R. "*Over Empires and Oceans: Pioneers, Aviators and Adventurers.*" East Sussex: Tattered Flag Press, 2014.

Tipton, Stewart G., "*Air Transport 1971.*" Washington D.C.: Air Transport Association of America, 1971.

Lawrence, M. "*A Big Life In Advertising.*" New York: Touchstone, 2002.

Davies, R.E.G. "*Airlines of the Jet Age: A History.*" Washington D.C.: Smithsonian, 2016.

Ganson, Barbara. "*Texas Takes Wing: A Century of Flight in the Lone Star State*." Austin: University of Texas Press, 2014.

Bleakley, Bruce A. "*Dallas/Fort Worth International Airport*." Charleston: Arcadia Publishing, 2013.

Huhne, M. C. "*Airline Visual Identity: 1945-1975.*" Berlin: Callisto Publishers, 2014.

Petzinger, Thomas. "*Hard Landing: The Epic Contest for Power and Profits that Plunged the Airlines Into Chaos*." New York: Times Business, 1996.

Nance, John J. "*Splash of Colors: The Self-destruction of Braniff International*." New York: Morrow, 1984.

Uslay, Can., Dixit, Ashutosh., Sheth, Jagdish N.., Allvine, Fred C. "*Deregulation and Competition: Lessons from the Airline Industry*." New Dehli: SAGE Publications, 2007.

Vantoch, Victoria. "*The Jet Sex: Airline Stewardesses and the Making of an American Icon*." Pennsylvania: University of Pennsylvania Press, Incorporated, 2013.

Smith, Myron J. "*The Airline Encyclopedia, 1909-2000*." Oxford: Scarecrow Press, 2002.

Leary, William M. "*The Airline Industry*." Oxford: Bruccoli Clark Layman Inc. 1992

DuBois, Arthur W. "*Behind the Facade and a Peek at Panagra*." Bloomington: Xlibris US, 2009.

Bailey, Elizabeth E., Graham, David P.., Kaplan, Daniel P.., Graham, David R. "*Deregulating the Airlines*." Massachusetts: MIT Press, 1985.

Szurovy, Geza. "*Classic American Airlines*." St. Paul: Zenith Press, 2003.

Yenne, Bill. "*Classic American Airliners*." St. Paul: Zenith Press, 2005.

Gilchrist, Peter. "*Boeing 747 Classic*." Wisconsin: MBI Publishing Company, 2000.

Skinner, Stephen. "*BAC One-Eleven: The Whole Story*." Cheltenham: The History Press, 2013.

Balch, Adrian. "*Boeing 727*." Shrewsbury: Airlife Publishing Ltd., 1992

Green, William., Swanborough, Gordon. "*Airlines and Airliners Directory*." London: Frederick Warne (Publishers) Ltd., 1980.

Other resources:

Lois, George., Pitts, Bill. "*Tales of Madison Avenue.*" New York Magazine. October 9, 1972.

Clay, Jennifer Coutts. "*Flying Colours*." Aircraft Interiors International. June, 2007.

Fallows, James. "*The Great Airline War: Flying the not-so-friendly skies of Texas.*" TexasMonthly. December, 1975.

Scherer, Ron. "*Why Braniff International's colorful career has ended.*" The Christian Science Monitor. May 14, 1982.

Cuff, Daniel F. "*MAJOR EXPANSION IN LATE 70's OVERWHELMED COLORFUL BRANIFF*." The New York Times. May 13, 1982.

Tolman, Keith. "*BRANIFF INTERNATIONAL AIRWAYS.*" Oklahoma Historical Society. Accessed April 20, 2020. https://www.okhistory.org/publications/enc/entry.php?entry=BR006.

Magers, Michael., "*Inside the Sisterhood of Braniff Flight Attendants.*" Conde Nast Traveler. Accessed April 16, 2020. https://www.cntraveler.com/story/inside-the-sisterhood-of-braniff-flight-attendants.

Sorene, Paul. "*Emilio Pucci Uniforms For Braniff International Airlines' Hostesses, 1965-73*." Flashbak. Accessed April 16, 2020. https://flashbak.com/emilio-pucci-uniforms-for-braniff-international-airlines-stripping-hostesses-1965-73-48891/.

Stretton, Richard. "*Braniff's Flying Colors 727s*." Yesterday's Airlines. Accessed April 23, 2020. https://www.yesterdaysairlines.com/airline-history-blog/braniffs-flying-colors-727s.

Cole, Rachel. "*Braniff and the End of the Plain Plane*." Northwestern University. Accessed April 23, 2020. https://sites.northwestern.edu/northwesternlibrary/2016/08/23/braniff-and-the-end-of-the-plain-plane/.

Dwyer-Lindgren, Jeremy. "*The Other Concorde Airlines: Braniff International & Singapore Airlines*." Airways Magazine. Accessed April 19, 2020. https://airwaysmag.com/uncategorized/concorde-airlines-braniff-international-singapore-airlines/.

Concorde Key Events. "*Braniff Airways - British Airways / Air France joint services 1979-1980*." Concorde Key Events. Accessed April 19, 2020. http://www.concordesst.com/history/events/braniff.html.

Heritage Concorde. "*Braniff Airways Concorde Operations*." Heritage Concorde. Accessed April 19, 2020. https://www.heritageconcorde.com/braniff-airways-concorde-operations.

INDEX

Made in the USA
Las Vegas, NV
03 November 2026

57424072R00059